Revision Notes

OCR GCSE
Mathematics A

Higher

Howard Baxter, Ruth Crookes,
Mike Handbury, John Jeskins,
Jean Matthews, Colin White

Series Editor: Brian Seager

- Personalise your revision
- Learn essential facts
- Practise exam questions
- Get your best grade!

 HODDER
EDUCATION

Acknowledgements

Every effort has been made to trace all copyright holders, but if any have been inadvertently overlooked the Publishers will be pleased to make the necessary arrangements at the first opportunity.

Although every effort has been made to ensure that website addresses are correct at time of going to press, Hodder Education cannot be held responsible for the content of any website mentioned in this book. It is sometimes possible to find a relocated web page by typing in the address of the home page for a website in the URL window of your browser.

Hachette UK's policy is to use papers that are natural, renewable and recyclable products and made from wood grown in sustainable forests. The logging and manufacturing processes are expected to conform to the environmental regulations of the country of origin.

Orders: please contact Bookpoint Ltd, 130 Milton Park, Abingdon, Oxon OX14 4SB. Telephone: (44) 01235 827720. Fax: (44) 01235 400454. Lines are open 9.00 – 5.00, Monday to Saturday, with a 24-hour message answering service. Visit our website at www.hoddereducation.co.uk

Howard Baxter, Ruth Crookes, Mike Handbury, John Jeskins, Jean Matthews, Mark Patmore, Brian Seager, Eddie Wilde, Colin White 2011

First published in 2011 by
Hodder Education,
An Hachette UK Company
338 Euston Road
London NW1 3BH

Impression number   5 4 3 2 1
Year   2015 2014 2013 2012 2011

Cover illustration by Stankevich/Shutterstock
Typeset in 10/12 Cronos Pro by Pantek Media, Maidstone, Kent
Printed in Spain

A catalogue record for this title is available from the British Library

ISBN: 978 1444 145854

Getting the most from this book

This book will help you prepare for the examinations in Units A, B and C of the Higher Tier of OCR GCSE Mathematics A (J562). It covers all you need to know for the examinations. The topics from the specification are in the same order and with the same chapter numbers as the Student's Book.

Each topic has been introduced in the same way: a reminder of definitions and techniques; a Test Yourself question, with the corresponding solution and frequently, a 'Chief Examiner says' tip to give further guidance and help you tackle examination questions. For each topic there is an exam question together with its solution available on a website that you can access via http://www.hodderplus.co.uk/myrevisionnotes. This is followed by further exam questions for you to attempt. The answers to all the extra questions are also on the website.

How to use this book

- Open a page randomly and check that you can do the questions, or else use the book systematically as part of your planned revision.

- Whether you choose to work through the topics in order or jump between topic areas, you can check on your progress using the Revision records at the front.

- Choose a topic, read the reminders and try the Test Yourself question. If you get it right, you can go on to the next question. If not, look at where you went wrong.

- At the end of the topic, go to the website and try the Exam question without looking at the solution. When you have understood that, do all the More Exam Practice questions. Finally, don't forget to tick the box on your Revision record.

- If you know that your knowledge is worse in certain topics, don't leave these to the end of your revision programme. Put them in at the start so that you have time to return to them nearer the end of the revision period.

Remember that you are **not** allowed to use a calculator in Unit B. You will need a calculator in Units A and C to answer some of the questions.

Examination tips

There will be a formulae sheet on the second page of the exam paper. It contains several formulae. Make sure you know what they are.

Read the instructions carefully, both those on the front of the paper and those in each question. Here are the meanings of some of the words used:

- **Write, write down, state** – little working out will be needed and no explanation is required.

- **Calculate, find** – something to be worked out, with a calculator if appropriate. It is a good idea to show the steps of your working as this may earn marks even if the answer is wrong.

- **Solve** – show all the steps in solving the equation.

- **Prove, show** – all the steps needed, including reasons, must be shown in a logical way.

- **Deduce, hence** – use a previous result to help you find the answer.

- **Draw** – draw as accurately and carefully as you can.

- **Sketch** – need not be accurate but should show the essential features.

- **Explain** – give (a) reason(s). The number of features you need to mention can be judged by looking at the marks. One mark probably means only one reason is required.

As well as this book, there are a lot of websites that will help you revise. Go to http://www.m-a.org.uk/links/revision for a list of links.

Brian Seager, 2011
Series Editor

Contents and revision planner

A Mathematics

B Mathematics

C Mathematics

Exam questions – online

Answers to exam questions – online

1 Working with numbers

Powers and roots on your calculator

Revised

Use

- $\boxed{x^2}$ for squares
- $\boxed{x^3}$ for cubes
- $\boxed{\sqrt{}}$ for square roots
- $\boxed{\sqrt[3]{}}$ for cube roots
- $\boxed{\wedge}$ or $\boxed{x^y}$ or $\boxed{y^x}$ or $\boxed{x^{\square}}$ for powers
- $\boxed{\sqrt[x]{}}$ or $\boxed{\sqrt[x]{y}}$ or $\boxed{y^{\frac{1}{x}}}$ or $\boxed{\sqrt[\square]{\square}}$ for roots.

You may need to use the $\boxed{\text{SHIFT}}$ or $\boxed{\text{INV}}$ or $\boxed{\text{2nd F}}$ key to use some of these functions.

Test Yourself 1

Work out these. Where necessary, give your answer correct to 2 decimal places.

a 6^7 **b** 3.7^4

c $\sqrt[3]{6.2}$ **d** $\sqrt[5]{29}$

Solutions

Test Yourself 1

a Key in $\boxed{6}\boxed{\wedge}\boxed{7}\boxed{=}$ 279 936

b Key in $\boxed{3}\boxed{.}\boxed{7}\boxed{\wedge}\boxed{4}\boxed{=}$ 187.4161

c Key in $\boxed{\sqrt[3]{}}\boxed{6}\boxed{.}\boxed{2}\boxed{=}$ 1.84 (2 d.p.)

d Key in $\boxed{5}\boxed{\sqrt[x]{}}\boxed{2}\boxed{9}\boxed{=}$ 1.96 (2 d.p.)

Reciprocals

Revised

- The reciprocal of x is $\dfrac{1}{x}$.
- The reciprocal of $\dfrac{a}{b}$ is $\dfrac{b}{a}$.
- Use the $\boxed{1/x}$ or the $\boxed{x^{-1}}$ key.

Test Yourself 2

Work out the reciprocal of each of these.

a 4   **b** $\frac{2}{3}$

c 0.1   **d** 40

Solutions

Test Yourself 2

a $\frac{1}{4}$ or 0.25

b $\frac{3}{2}$ or $1\frac{1}{2}$

> You shouldn't need your calculators for these.

c Key in $\boxed{0}\boxed{.}\boxed{1}\boxed{1/x}\boxed{=}$ 10

d Key in $\boxed{4}\boxed{0}\boxed{1/x}\boxed{=}$ 0.025

Fractions on your calculator

- To input fractions, use the $a\frac{b}{c}$ button.
  For example, to input $2\frac{3}{4}$, key in ② $a\frac{b}{c}$ ③ $a\frac{b}{c}$ ④.

Solutions

Test Yourself 3

a Key in ② ⑤ $a\frac{b}{c}$ ① ⓪ ⓪ ＝

  Display = ⌐1 ⌐ 4⌐ = $\frac{1}{4}$

b Key in

  ① $a\frac{b}{c}$ ② $a\frac{b}{c}$ ③ ＋ ③ $a\frac{b}{c}$ ④ ＝

  Display = ⌐2 ⌐ 5 ⌐ 12⌐

  $= 2\frac{5}{12}$

Test Yourself 3

a Write $\frac{25}{100}$ as a fraction in its simplest form.

b Work out $1\frac{2}{3} + \frac{3}{4}$.

Order of operations

- When doing complex calculations with a calculator, you need to use brackets. For example, to work out

  $\dfrac{3.6 + 2.7}{1.4 + 5.2}$, you need to key in ⦅ ③ ⦁ ⑥ ＋ ② ⦁ ⑦ ⦆

  ÷ ⦅ ① ⦁ ④ ＋ ⑤ ⦁ ② ⦆ ＝

- Multiplication and division must be carried out before addition and subtraction. For example, to calculate, $3 + 4 \times 5$, you must work out 4×5 first. A calculator does this automatically – you can just key in

  ③ ＋ ④ × ⑤ ＝

  and it will do the calculation in the correct order.

- Brackets must be worked out before multiplication and division. For example, to calculate $(3 + 4) \times 5$, you must work out $3 + 4$ first. If you key the calculation, including the brackets, into a calculator, it will do the whole calculation, in one go.

Chief Examiner says

Remember that for every ⦅ there must be a ⦆.

Test Yourself 4

a Work out $\dfrac{4.2 + 3.6}{1.7}$ correct to 2 decimal places.

b Work out $4.7 \times 3.2 - 1.7 \times 6.1$

Solutions

Test Yourself 4

a ⦅ ④ ⦁ ② ＋ ③ ⦁ ⑥ ⦆ ÷ ① ⦁ ⑦ ＝ 4.588 235

  ... = 4.59 (2 d.p.)

b ④ ⦁ ⑦ × ③ ⦁ ② － ① ⦁ ⑦ ×

  ⑥ ⦁ ① ＝ 4.67

Chief Examiner says

It is easy to press the wrong key when using a calculator. Do the calculation twice. Also, check if your answer seems sensible and find an estimate by approximating.

Rounding to a given number of decimal places

- If the first digit to be rounded is 4 or less, then just ignore it. For example, 5.1246 correct to 2 decimal places is 5.12.

  > The first digit to be rounded is 4, so write down the required digits.

- If the first digit to be rounded is 5 or more, then add one to the last required digit. For example, 6.2873 correct to 2 decimal places is 6.29.

  > The first digit to be rounded is 7, so add 1 to the last required digit, 8.

Solutions

Test Yourself 5

a 0.2 — The next digit is 3, so leave 2.

b 21.4 — The next digit is 6, so add 1 to 3.

c 14.2 — The next digit is 9, so add 1 to 1.

d 2.4 — The next digit is 0, so leave 4.

e 244.0 — The next digit is 5, so add 1 to 9, which changes 3.9 to 4.0. If the last digit required is zero you must include it.

Rounding to a given number of significant figures

- In any number, the first digit from the left which is not 0 is the first significant figure.
- When rounding to a given number of significant figures,
  - if the first digit to be rounded is 4 or less, the just ignore it and, if necessary, include the correct number of zeros to show the size of the number. For example, 0.213 to 1 significant figure is 0.2 and 213 to 1 significant figure is 200.
  - if the first digit to be rounded is 5 or more, add 1 to the last significant figure and, if necessary, include the correct number of zeros to show the size of the number. For example, 0.078 to 1 significant figure is 0.08 and 78 to 1 significant figure is 80.

> **Chief Examiner says**
>
> The number 0.96 correct to 1 significant figure is 1, not 1.0, which has 2 significant figures. Extra zeros are used only to show the size of the number, as in 200, for example.

Solutions

Test Yourself 6

a 4

b 0.05

c 900

d 400 000

Prime numbers and factors and Writing a number as a product of its prime factors

Revised ☐

- Prime numbers are numbers that have only two factors, 1 and the number itself.
- Prime factors are factors that are also prime numbers.
- You need to be able to write numbers as the product of their prime factors. For example, $12 = 2 \times 2 \times 3 = 2^2 \times 3$.
- There are different methods of finding the product of the prime factors. Use whichever method you prefer.

Test Yourself 7

a Write down all the prime numbers between 40 and 50.

b Write 48 as the product of its prime factors.

Solutions

Test Yourself 7

a 41, 43, 47

b Method 1

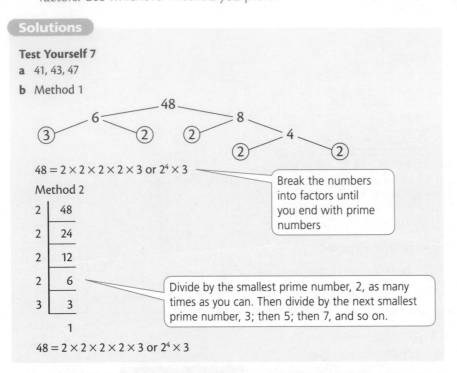

$48 = 2 \times 2 \times 2 \times 2 \times 3$ or $2^4 \times 3$

> Break the numbers into factors until you end with prime numbers

Method 2

2	48
2	24
2	12
2	6
3	3
	1

> Divide by the smallest prime number, 2, as many times as you can. Then divide by the next smallest prime number, 3; then 5; then 7, and so on.

$48 = 2 \times 2 \times 2 \times 2 \times 3$ or $2^4 \times 3$

Highest common factors and lowest common multiples

Revised ☐

Highest common factor (HCF)

- The HCF of a set of numbers is the largest number that will divide into each of the numbers.
- To find the HCF, write each number as the product of its prime factors and choose the factors that are common to all the given numbers.

Test Yourself 8

Find the HCF of 120, 36 and 84.

Solution

Test Yourself 8

$120 = 2 \times 2 \times 2 \times 3 \times 5 = 2^3 \times 3 \times 5$

$36 = 2 \times 2 \times 3 \times 3 = 2^2 \times 3^2$

$84 = 2 \times 2 \times 3 \times 7 = 2^2 \times 3 \times 7$

HCF is $2^2 \times 3 = 12$

> 2 and 3 are common to all three numbers. The lowest power of 2 is 2^2 and of 3 is 3^1.

Chief Examiner says

Check that 12 divides into 120, 36 and 84.

Lowest common mulitple (LCM)

- The LCM of a set of numbers is the smallest number that can be divided exactly by all the numbers. For example, the LCM of 8 and 6 is 24.

- To find the LCM, write each number as the product of its prime factors and choose the largest power of each prime that appears in the lists.

Test Yourself 9

Find the LCM of 120, 36 and 84.

Solution

Test Yourself 9

$120 = 2 \times 2 \times 2 \times 3 \times 5 = 2^3 \times 3 \times 5$

$36 = 2 \times 2 \times 3 \times 3 = 2^2 \times 3^2$

$84 = 2 \times 2 \times 3 \times 7 = 2^2 \times 3 \times 7$

LCM is $2^3 \times 3^2 \times 5 \times 7 = 2520$

> 2, 3, 5 and 7 appear in these lists. The highest power of 2 is 2^3, of 3 is 3^2, of 5 is 5^1 and of 7 is 7^1.

Chief Examiner says

Check that 120, 36 and 84 all divide into 2520.

Multiplying and dividing by negative numbers

Revised

- When multiplying and dividing positive and negative numbers, use these rules:
  - $+ \times + = +$
  - $+ \times - = -$
  - $- \times + = -$
  - $- \times - = +$
  - $+ \div + = +$
  - $+ \div - = -$
  - $- \div + = -$
  - $- \div - = +$

- Combine the numbers and signs separately.

- To input negative numbers on a calculator, use $(-)$ or $+/-$.

Test Yourself 10

a Work out these without your calculator.

i -4×3

ii $-12 \div -6$

b Use your calculator to work out these.

i $2 \times (-5)$

ii $-3 \div -3$

Chief Examiner says

To input −5 on some older calculators you need to input 5 $(-)$ rather than $(-)$ 5. Make sure you know which is needed on your calculator.

Solutions

Test Yourself 10

a i $-4 \times 3 = -12$ — $- \times + = -$

ii $-12 \times -6 = 2$ — $- \div - = +$

b i Key in

2 $\times$ $(-)$ 5 $=$ -10

ii Key in

$(-)$ 3 $\div$ $(-)$ 3 $=$ 1

2 Algebra

Expanding brackets and Collecting like terms
Revised

- When expanding brackets, multiply every term within the brackets by the term immediately in front of the bracket.
- Like terms can be added and subtracted.

Test Yourself 1

a Expand $3x(x-2)$.

b Expand and simplify $3(2a+3b)-2(a-2b)$.

Solutions

Test Yourself 1

a $3x^2-6x$

b $6a+9b-2a+4b=4a+13b$ — Collect like terms.

Factorising
Revised

- Look for every number and letter that is common to every term and write these outside the bracket.
- Write the terms in the bracket needed to give the original expression.

Test Yourself 2

Factorise the following.

a $3a+6$

b $2x^2-3xy$

c $4a-8ac+16a^2b$

Solutions

Test Yourself 2

a $3(a+2)$

b $x(2x-3y)$

c $4a(1-2c+4ab)$ — Note that $4a\times1=4a$ so 1 must be in the bracket.

> **Chief Examiner says**
>
> Always try to take out as big a factor as possible. In part **c)** 2 is a common factor, 4 is a common factor, and a is a common factor. The biggest common factor is $4a$.

Index notation
Revised

- You can use index notation in algebra.
- $x\times x\times x\times x=x^4$

Test Yourself 3

Write these using index notation.

a $b\times b\times b$

b $x\times x\times y\times y\times y$

c $2x\times3x$

Solutions

Test Yourself 3

a b^3

b x^2y^3

c $6x^2$ — Multiply the numbers together, then multiply the letters together.

3 Statistical diagrams

Drawing and interpreting pie charts

- To draw a pie chart,
  - first calculate the angle that represents one observation by dividing 360° by the total frequency
  - then multiply this amount by the frequency for each group.

Solutions

Test Yourself 1

a 360° ÷ 30 = 12°, so one student is represented by 12°.

Calculate the angle for each sector.

Activity	Frequency
Ice skating	6 × 12 = 72°
Paint balling	8 × 12 = 96°
Horse riding	5 × 12 = 60°
Theme park	11 × 12 = 132°
Total	360°

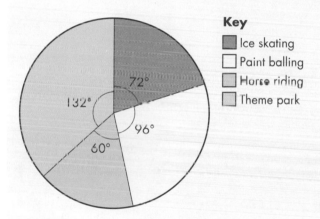

Key
- Ice skating
- Paint balling
- Horse riding
- Theme park

b 60° represents 20 students.

$$\frac{60°}{360°} = \frac{1}{6}$$

so the total number of students = 20 × 6 = 120

Test Yourself 1

Class 10h had a choice of four events for their activity day.

The table shows the number who chose each activity.

Activity	Frequency
Ice skating	6
Paint balling	8
Horse riding	5
Theme park	11
Total	30

a Draw a pie chart to show this information.

b Look at the pie chart you have drawn.

If you were told the number of students choosing horse riding was 20, what would have been the total number of students?

Chief Examiner says

Don't forget to label the sectors with the groups, not the size of the angles. The angles are labelled here as a check for you.

Frequency diagrams and Frequency polygons

- For grouped information with equal intervals, a frequency diagram can be drawn with the heights of the bars equal to the frequency.
- A frequency polygon is formed by joining, with straight lines, the midpoints of the tops of the bars in a frequency diagram.

Test Yourself 2

The table shows the amount of pocket money received by the members of Class 10h.

To show this information, draw the following.

a A frequency diagram

b A frequency polygon

Amount (£)	0–3.99	4–7.99	8–11.99	12–15.99
Frequency	3	10	12	5

Solutions

Test Yourself 2

a

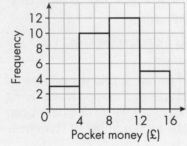

b

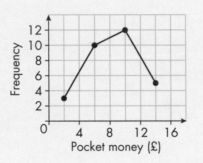

Stem-and-leaf-diagrams

- You can display a large number of two-digit data items in a stem-and-leaf diagram.
- You write the tens digit (the stem) in the left-hand column.
- You write the units digits (the leaves) in the right-hand column.
- To draw a stem-and-leaf diagram,
  - first work through the data and write the 'leaf' for each value in turn
  - then redraw the diagram with the 'leaves' ordered.
- You can also use a stem-and-leaf diagram with, for example, units as the 'stem' and the first decimal place as the 'leaves'.
- Always give a key, for example 3 | 2 represents 32.
- You can find the median by counting up to the middle value.

Comparing distributions

- A back-to-back stem-and-leaf diagram can be used to compare two sets of data. To compare the two sets of data state which set of data
  - is bigger on average using the median
  - is more spread out using the range.

Test Yourself 3

Here are the test marks for a group of students.

63	54	34	42
38	76	69	52
45	54	67	71
64	43	57	59
47	66	57	63

a Draw a stem-and-leaf diagram for these results. Don't forget the key.

b Use your stem-and-leaf diagram to find the median score.

Test Yourself 3

a Unordered

3	4	8			
4	2	5	3	7	
5	4	2	4	7	9	7
6	3	9	7	4	6	3
7	6	1			

Ordered

3	4	8			
4	2	3	5	7	
5	2	4	4	7	7	9
6	3	3	4	6	7	9
7	1	6			

Key: 3 | 4 represents 34

b Since there are 20 values there are two middle values, the 10th and 11th.
They are both 57, so the median is 57.

Test Yourself 4

The back-to-back stem-and-leaf diagram below shows the marks scored by Classes 10A and 10B in their last Maths test.

```
          10A                        10B
                          | 1 | 1 8
                    7 6 5 | 2 | 2 5
              8 6 4 3 2 2 | 3 | 3 6 7
        9 8 7 6 6 3 1 0 | 4 | 2 5 5 6 7 8
            9 8 6 4 3 | 5 | 4 6 6 7 8 9 9
            6 5 4 4 | 6 | 0 2 3 5 6 8
              1 1 0 | 7 | 1 3 4
```

Key: 3 | 2 represents 32
Make two comparisons between the two distributions.

Solution

Test Yourself 4

10A: Median (15th mark) = 47   Range = 71 − 25 = 46

10B: Median (15th mark) = 56   Range = 74 − 11 = 63

So Class 10B have higher scores on average (median) and also have a greater spread of scores (range).

4 Equations

Solving equations

- When you solve an equation, you must always do each operation to the whole of both sides of the equation.
- An equation with an x^2 term usually has two solutions.

Chief Examiner says

Remember that $5x^2$ means $5 \times x^2$ not $(5x)^2$.

Test Yourself 1

Solve these equations.

a $3x - 4 = 20$

b $5x^2 = 80$

Solutions

Test Yourself 1

a $3x - 4 = 20$

$3x = 24$ ← First add 4 to both sides.

$x = 8$ ← Then divide both sides by 3.

b $5x^2 = 80$

$x^2 = 16$ ← First divide both sides by 5.

$x = \pm 4$ ← Then find the square root of each side.

Solving equations with brackets

- If an equation has a bracket, multiply out the bracket first.
- If there is more than one bracket,
  - multiply out all the brackets first
  - then simplify both sides before rearranging.

Test Yourself 2

Solve

a $3(x + 4) = 5$

b $5(a + 2) - 3(a + 1) = 8$

Solutions

Test Yourself 2

a $3(x + 4) = 5$

$3x + 12 = 5$

$3x = -7$

$x = -\frac{7}{3}$

b $5(a + 2) - 3(a + 1) = 8$

$5a + 10 - 3a - 3 = 8$ ← $-3 \times 1 = -3$

$2a + 7 = 8$

$2a = 1$

$a = \frac{1}{2}$

Chief Examiner says

With an equation such as the one in part **b)** it is a common error to get the sign in front of the 3 wrong.

Equations with *x* on both sides

Revised

- If the unknown quantity is on both sides of the equation, rearrange the equation to collect the numbers on one side and the unknown on the other side.

- It is usually easier to move the unknown to the side where the coefficient (number in front) is positive.

Test Yourself 3

Solve

$$5x + 4 = 2x + 19$$

Solution

Test Yourself 3

$5x + 4 = 2x + 19$

$5x = 2x + 15$ ——— (Subtract 4 from both sides.)

$3x = 15$ ——— (Subtract $2x$ from both sides.)

$x = 5$

Fractions in equations

Revised

- If the equation has a fraction in it, multiply by the denominator.

Test Yourself 4

Solve $\dfrac{2(x+2)}{3} = 4$

Solution

Test Yourself 4

$\dfrac{2(x+2)}{3} = 4$

$2(x+2) = 12$ ——— (Multiply both sides by 3.)

$2x + 4 = 12$

$2x = 8$

$x = 4$

5 Ratio and proportion

Revised

What is a ratio? and Writing a ratio in the form 1 : n

- If two quantities are in proportion to one another so that for x parts of the first there are y parts of the second, then the ratio is x to y, which is written as $x : y$.
- To write a ratio in its simplest form, both parts need to be in the same units and then divided by the same number until they have no common factors.
- To write a ratio in the form 1 : n, divide the second part by the first.

Test Yourself 1

Write each of these ratios

i in its lowest terms

ii in the form 1 : n.

a 6 : 9

b 50p : £4

c 5 m : 40 cm

Solutions

Test Yourself 1

a i 2 : 3

Divide both sides by 3.

ii 1 : 1.5

Divide both sides by 2.

b i 50 : 400 = 1 : 8

Change to pence and divide both sides by 50.

ii 1 : 8

c i 500 : 40 = 25 : 2

Change to cm and divide both sides by 20.

ii $1 : \frac{2}{25} = 1 : 0.08$

Divide both sides by 25.

Using ratios

Revised

- If two substances are in the ratio 1 : n,
  - to find the second amount from the first, the first must be multiplied by n
  - to find the first amount from the second, the second must be divided by n.
- If two substances are in the ratio m : n,
  - to find the second from the first, the first must be multiplied by $\frac{n}{m}$
  - to find the first from the second, the second must be multiplied by $\frac{m}{n}$.

Test Yourself 2

In a recipe, flour and fat are used in the ratio 5 : 2.

a How much fat is needed when 480 g of flour is used?

b How much flour is needed when 300 g of fat is used?

Solutions

Test Yourself 2

a Quantity of fat $= 480 \times \frac{2}{5}$

$= 192$ g

b Quantity of flour $= 300 \times \frac{5}{2}$

$= 750$ g

Dividing a quantity in a given ratio

Revised

- To share in a given ratio, first add the parts of the ratio together. Divide the amount to be shared by this total. This is then the multiplier for each of the parts.
- If the amount to be shared is not given, the multiplier can be found by dividing the part given by its part of the ratio. This can then be used to find the other parts or the total.

Chief Examiner says

If the multiplier is not an exact amount, leave it as a fraction and round the final answer if necessary.

Test Yourself 3

£150 is shared in the ratio 2 : 3 : 5. How much is each part?

Test Yourself 4

In a school students must study one foreign language but cannot study more than one.

Students from one year group chose French, Spanish, Latin or German in the ratio 6 : 5 : 2 : 3.

42 students study French.

How many students are there in the year group?

Solutions

Test Yourself 3
Total parts is $2 + 3 + 5 = 10$
Multiplier is $150 \div 10 = 15$
Parts are $2 \times 15 = £30$
$3 \times 15 = £45$, $5 \times 15 = £75$

Chief Examiner says

Check that
$30 + 45 + 75 = 150$

Test Yourself 4
Multiplier $= 42 \div 6 = 7$

Students studying French ÷ French ratio

Total ratio $= 6 + 5 + 2 + 3 = 16$
Total number of students
$16 \times 7 = 112$

Best value

Revised

You need to be able to

- assess value for money, by working out and comparing cost per unit.
- deal with other ratio problems.
- calculate currency exchange, by multiplying or dividing by an exchange rate.

Chief Examiner says

In a question that asks for the best value, if you show no working, you will score no marks, even if you pick the correct item.

Test Yourself 5

Semi-skimmed milk is sold at these prices:
500 ml for 40p; 1 litre for 78p; 2 pints (1.136 litre) for 86p.
Which is the best buy?

Solution

Test Yourself 5
500 ml costs $40 \times 2 = 80$p a litre.
1 litre costs 78p a litre.
2 pints costs $86 \div 1.136 = 75.7$p a litre.
So the 2-pint size is the best value.

6 Statistical calculations

The mean from a frequency table

Revised

- To calculate the mean from a frequency table,
  - multiply each observation (x) by its corresponding frequency (f)
  - then add up these results
  - finally divide by the total frequency.
- You can add an extra row or column to the table to calculate $x \times f$.

> **Chief Examiner says**
>
> You are often told the total frequency in the question.

Test Yourself 1

The table shows the number of letters delivered one morning to a street of 100 houses. Find the mean number of letters.

No. of letters (*x*)	0	1	2	3	4	5
Frequency	8	19	28	25	17	3

Solution

Test Yourself 1

No. of letters (*x*)	0	1	2	3	4	5	Total
Frequency (*f*)	8	19	28	25	17	3	100
x × *f*	0	19	56	75	68	15	233

Mean = 233 ÷ 100 = 2.33

> **Chief Examiner says**
>
> Check that the total frequency matches that given in the question.

Grouped data and Continuous data

Revised

- For grouped discrete data, use the middle of the interval for the value of x to represent each group. For example, the middle value of the group 0–9 is 4.5.
- For continuous data, again, use the middle of the interval for the value of x to represent each class. For example, the middle value of the class $0 \leqslant x < 10$ is 5.

Test Yourself 2

The table shows the amount of pocket money received by the members of Class 10h.

Amount (£)	0–3.99	4–7.99	8–11.99	12–15.99
Frequency	3	10	12	5

Calculate an estimate of the mean amount of pocket money received by these students.

Solution

Test Yourself 2

Amount (£x)	0–3.99	4–7.99	8–11.99	12–15.99	Total
Frequency (f)	3	10	12	5	30
Middle (x) value	2	6	10	14	
f × middle x	6	60	120	70	256

Estimate of the mean = 256 ÷ 30 = £8.53 (2 d.p.)

Test Yourself 3

A gardener picked some runner beans and measured their lengths. The table shows the results.

Length of bean in centimetres (x)	Frequency (f)
$10 \leqslant x < 15$	3
$15 \leqslant x < 20$	10
$20 \leqslant x < 25$	26
$25 \leqslant x < 30$	5
$30 \leqslant x < 35$	1

Calculate an estimate of the mean.

Solution

Test Yourself 3

Length of bean in centimetres (x)	Frequency (f)	Middle (x) value	f × middle x
$10 \leqslant x < 15$	3	12.5	37.5
$15 \leqslant x < 20$	10	17.5	175
$20 \leqslant x < 25$	26	22.5	585
$25 \leqslant x < 30$	5	27.5	137.5
$30 \leqslant x < 35$	1	32.5	32.5
	45		967.5

Estimate of mean = 967.5 ÷ 45 = 21.5 cm

7 Pythagoras' theorem

Pythagoras' theorem and Using Pythagoras' theorem
Revised ☐

- In a right-angled triangle, labelled like this, Pythagoras' theorem states that $a^2 = b^2 + c^2$.
- The longest side of a right-angled triangle is called the hypotenuse.

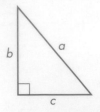

Chief Examiner says
- It is a good idea to label the triangles, using letters for the sides, and to write down the rule; then substitute the actual values for the sides before rearranging, if that is needed.
- As a check, remember that in a right-angled triangle the longest side, the hypotenuse, is always opposite the right-angle.

Solutions

Test Yourself 1

a $a^2 = b^2 + c^2$ ⟵ *a is the hypotenuse.*

$= 8^2 + 6^2$

$= 64 + 36$

$= 100$

$a = \sqrt{100}$

$= 10\,\text{cm}$

b $a^2 = b^2 + c^2$

$8^2 = b^2 + 4^2$

$64 = b^2 + 16$

$b^2 = 64 - 16$

$b^2 = 48$

$b = \sqrt{48}$

$b = 6.9\,\text{cm}$ (1 d.p.)

Test Yourself 1
Calculate the length of the unknown side in each of these triangles.

a

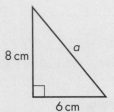

8 cm a

6 cm

b

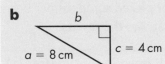

b

$a = 8\,\text{cm}$ $c = 4\,\text{cm}$

Pythagorean triples
Revised ☐

- Three positive integers (a, b, c) form a Pythagorean triple if $a^2 + b^2 = c^2$.
- If the lengths of three sides of a triangle form a Pythagorean triple, then the triangle is right-angled.

Solutions

Test Yourself 2

a   Yes $81 + 1600 = 1681 = 41^2$

b   Yes $30.25 + 900 = 930.25 = 30.5^2$

The Pythagorean triple is double this: 11, 60, 61.

c   No $400 + 441 = 841$ but $28^2 = 784$

Test Yourself 2
Are these right-angled triangles?

a

41

9

40

b

30.5

5.5

30

c

20

21

28

Using Pythagoras' theorem in three dimensions

- You can find lengths in three-dimensional objects by identifying right-angled triangles within the object.

- To find the length of the diagonal of a cuboid, use Pythagoras' theorem to find the diagonal of a face of the cuboid, then use it again to find the diagonal of the whole cuboid.

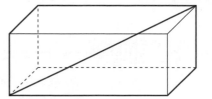

Solution

Test Yourself 3

The length of the diagonal of face $= \sqrt{15^2 + 8^2} = \sqrt{289}$

(Don't work this out because you are going to square it in the next step.)

The length of the diagonal $= \sqrt{289 + 6^2} = \sqrt{325}$

$\qquad\qquad\qquad\quad = 18.0\,\text{cm}$ (to 1 d.p.)

Test Yourself 3

Calculate the length of the diagonal of a cuboid measuring 15 cm by 8 cm by 6 cm.

Line segments

- A line segment is the part of a line between two points. It has a finite length.

- To find the midpoint of the line segment joining A and B, find the mean of their coordinates.

- To find the length of the line segment joining two points, use Pythagoras' theorem. The difference in the x coordinates and the difference in the y coordinates give you the lengths of the two shorter sides.

Test Yourself 4

A is the point (1, 5) and B is the point (7, 2).

Calculate

a the midpoint of AB

b the length of AB.

Solutions

Test Yourself 4

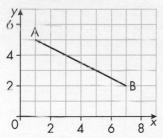

a The midpoint of AB is $\left(\dfrac{1+7}{2}, \dfrac{5+2}{2}\right) = (4, 3.5)$

b $AB^2 = 6^2 + 3^2 = 45$

$\quad AB = \sqrt{45} = 6.7$ units (1 d.p.)

Chief Examiner says

You don't have to draw a diagram but it may help.

8 Formulae 1

Using formulae

- When substituting in formulae, remember
  - an expression like ab means $a \times b$
  - multiplication and division are done before addition and subtraction unless brackets tell you otherwise
  - expressions like $3r^2$ mean $3 \times r^2$, that is you square r and then multiply by 3.
- When writing your own formulae,
  - say what the letters you use mean and include units
  - use brackets if you want addition and subtraction done before multiplication and division
  - use a fraction line not a $\div$ sign for divide.

> **Chief Examiner says**
>
> When substituting numbers in formulae, take care with the negative numbers.

Test Yourself 1

> Don't forget $3a^2$ means $3 \times a^2$.

a If $C = 6b + 3a^2$, find C when
  i $b = -3$ and $a = 5$
  ii $b = 4$ and $a = -2$

b If $y = \dfrac{3a - 2b}{c}$, find y when
  i $a = 2$, $b = -5$ and $c = 3$
  ii $a = \frac{1}{4}$, $b = \frac{3}{4}$ and $c = 2$

Test Yourself 2

Write a formula for the cost, £C, of m pens at £1.25 each and n pencils at 13p each.

Solutions

Test Yourself 1

a i $\quad C = 6 \times -3 + 3 \times 5^2$
   $\quad = -18 + 75 = 57$

> Don't forget that $(-2)^2 = +4$.

ii $\quad C = 6 \times 4 + 3 \times (-2)^2$
   $\quad = 24 + 12 = 36$

b i $\quad y = \dfrac{3 \times 2 - 2 \times (-5)}{3} = \dfrac{6 + 10}{3} = 5\frac{1}{3}$

ii $\quad y = \dfrac{3 \times \frac{1}{4} - 2 \times \frac{3}{4}}{2} = \dfrac{\frac{3}{4} - \frac{6}{4}}{2} = \dfrac{-\frac{3}{4}}{2} = -\frac{3}{8}$

Test Yourself 2
$C = 1.25m + 0.13n$

> Change 13p into £ so that all units are in £.

Rearranging formulae

- To change the subject of a formula, use the equation rule of doing the same to both sides to get the new subject on one side of the formula.

> **Chief Examiner says**
>
> Remember: always do the same operation to both sides.

Test Yourself 3

Make a the subject of $v = u + at$

Test Yourself 3

$v = u + at$

$v - u = at$ ── Take u from both sides.

$\dfrac{v - u}{t} = a$ ── Divide both sides by t. The fraction line acts as a bracket.

$a = \dfrac{v - u}{t}$ ── Rewrite with a on the left.

The language of algebra   Revised

- You need to be able to distinguish between formulae, expressions, equations and identities.
- Formulae have at least two unknown quantities. Values can be substituted to find an unknown. For example, $C = 5n + 2$.
- Expressions have no equals sign. For example, $y^2 - 6y$.
- Equations can be solved. For example, $2a + 7 = 12$.
- Identities are true for all values of the variable and so cannot be solved. Each side of an identity is a different way of expressing the other. For example, $2a + 7 = 3a + 2 - a + 5$.
- In an identity ≡ is sometimes used instead of =. For example, $2c + 6 \equiv 2(c + 3)$

Test Yourself 4

State whether each of the following is a formula, an expression, an equation or an identity.

a $3x - 5 = 4x + 2$

b $3y(y + 5)$

c $A = 2m(m + h)$

d $3y(y + 5) = 3y^2 + 15y$

e $4x - 13 = 7(x - 1) - 3(x + 2)$

Solutions

Test Yourself 4

a Equation ── The solution is $x = 7$.

b Expression ── There is no equals sign.

c Formula ── You could find A given the values of m and h.

d Identity ── If you work out the brackets on the left-hand side, you get the right-hand side.

e Identity ── If you expand the right-hand side you get $7x - 7 - 3x - 6$, which simplifies to $4x - 13$.

Chief Examiner says

Take care with the signs when expanding brackets.

9 Measures

Converting between metric units

- You need to know how to change between the common metric units and how to work with a mixture of units.

  - Length
    1 cm = 10 mm
    1 m = 1000 mm
    1 m = 100 cm
    1 km = 1000 m

  - Mass
    1 g = 1000 mg
    1 kg = 1000 g
    1 tonne = 1000 kg

  - Capacity/Volume
    1 litre = 1000 ml
    1 litre = 1000 cm³

Test Yourself 1

Change these units.

a 20 cm to mm

b 15 m to cm

c 2500 m to km

d 1.5 kg to g

e 4 litres to ml

Solutions

Test Yourself 1

a 20 cm = 20 × 10   [1 cm = 10 mm]
    = 200 mm

b 15 m = 15 × 100   [1 m = 100 cm]
    = 1500 cm

c 2500 m = 2500 ÷ 1000   [1 km = 1000 m]
    = 2.5 km

d 1.5 kg = 1.5 × 1000   [1 kg = 1000 g]
    = 1500 g

e 4 litres = 4 × 1000   [1 litre = 1000 ml]
    = 4000 ml

Converting between metric and imperial units

- These are the approximate equivalents between imperial and metric units.

  - 1 km is about $\frac{5}{8}$ mile or 1 mile is about 1.6 km.

  - 1 m is about 40 inches.

  - 1 inch is about $2\frac{1}{2}$ cm.

  - 1 foot is about 30 cm.

  - 1 kg is about 2 pounds.

  - 1 litre is about $1\frac{3}{4}$ pints.

  - 1 gallon is about $4\frac{1}{2}$ litres.   [You will only need to find approximate equivalents.]

- A more accurate approximation often used is 1 kg is about 2.2 pounds.

Test Yourself 2

Change these imperial measures to rough equivalent metric measures.

a 30 miles

b 15 pounds

c 10 pints

Solutions

Test Yourself 2

a 30 miles = 30 × 1.6   [1 mile = 1.6 km]
    = 48 km

b 15 pounds = 15 ÷ 2   [2 pounds = 1 kg]
    = 7.5 kg

c 10 pints = 10 ÷ 1.75   [$1\frac{3}{4}$ pints = 1 litre]
    = 6 litres (to the nearest litre)

Estimating measures

Revised

- You need to use the appropriate units when measuring. For example, somebody's height is measured in metres or centimetres, rather than millimetres or kilometres.

- It is useful to know some common measures for comparison when you are asked to estimate.
  - The height of a man is about 1.8 m.
  - The mass of a bag of sugar is about 1 kg.
  - Cans of drink hold about 300 ml.

Chief Examiner says

Do not try to be too accurate. For example, the length of a car is about 4 m, but 3 m or 5 m would also be accepted if you are not given more information.

Test Yourself 3

Estimate these in metric units.

a The height of your kitchen

b The length of your longest finger

c The mass of a bag of shopping

d The capacity of a washing-up bowl

Solutions

Test Yourself 3

a 3 m — Anything from 2.5 m to 4 m is acceptable, or 5 m if your kitchen is unusually high.

b 60 mm or 6 cm — Anything from 4 cm to 9 cm is acceptable.

c 5 kg — Anything from 3 kg to 8 kg is acceptable.

d 8 litres — Anything from 5 litres to 10 litres is acceptable.

Bearings and scale drawings

Revised

- Bearings are measured clockwise from North and must have three figures. An angle of 27° would be written as a bearing of 027°.

- The bearing of B from A is the angle at A, clockwise from North to the line to B marked x in the diagram. The bearing of A from B is the angle marked y in the diagram.

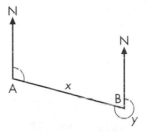

- Scales are sometimes expressed as ratios. For example, a scale of 1 : 10 000 on a map means that 1 cm on the map represents 10 000 cm (which is 100 m) on the ground.

Chief Examiner says

A useful rule to remember is that the difference between the bearing of B from A and the bearing of A from B is 180°.

Test Yourself 4

Measure the bearings of A, B and C from O.

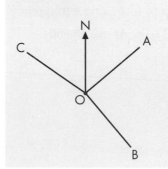

Solution

Test Yourself 4

A from O: 050°

B from O: 140°

C from O: 305°

10 Planning and collecting

Different types of data

- Primary data is data which is collected by the person using the information.
- Secondary data is data obtained from elsewhere such as the internet, books or databases.

Solutions

Test Yourself 1
a Secondary
b Primary

Test Yourself 1

State which type of data you would use in each of these cases.

a You are doing a project comparing the crime statistics in various parts of the UK.

b You are carrying out a survey on traffic around your school.

Data samples

- See also Chapter 13.
- When handling lots of data it is useful to collect the data in groups.
- Often grouped data is put into 'equal class intervals' – that is into groups of equal width.
- As a rough guide 5 to 10 groups is usually about right. If in doubt use smaller intervals to start with as you can always combine intervals later.
- Make sure your intervals do not overlap.

Test Yourself 2

Here are the times in minutes that 30 children claim to have spent on their homework.

26	48	52	23	21	59	29	31
26	33	23	33	37	44	29	26
21	45	31	60	52	51	39	54
37	39	40	52	39	48		

a Use tally marks to draw up a frequency table using intervals
   21–25, 26–30, 31–35, 36–40, 41–45, 46–50, 51–55, 56–60.

b Regroup to form a new frequency table using intervals.
   21–30, 31–40, 41–50, 51–60.

Test Yourself 2

a

Time (min)	Tally	Frequency
21–25	IIII	4
26–30	⊬↑↑	5
31–35	IIII	4
36–40	⊬↑↑ I	6
41–45	II	2
46–50	II	2
51–55	⊬↑↑	5
56–60	II	2

b

Time (min)	Frequency
21–30	9
31–40	10
41–50	4
51–60	7

Designing a questionnaire

Revised

- Questions you ask should not be biased nor should they be leading questions.
- Questions should give a choice of answers. This avoids a large number of vague answers that are difficult to analyse.
- Questions should not cause embarrassment.
- Questions should be short, clear, easy to understand and relevant.
- You should ask a variety of people at a variety of times in a variety of places so that your sample of people is unbiased – that is representative of the population as a whole.

Test Yourself 3

Criticise this question in a survey.

'Normal people enjoy watching *EastEnders*.

Do you watch *EastEnders*?'

Solution

Test Yourself 3

This is a biased question because if you do not watch 'soaps' in general or *EastEnders* in particular, it does not mean that you are not normal.

11 Sequences

Using rules to find terms of a sequence

Term-to-term rules

- The number of lines needed to make each of these pictures is

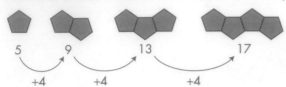

5 → 9 → 13 → 17
+4   +4   +4

- It is easy to see that the rule connecting the terms is +4. The next picture in the sequence would have 21 lines (17 + 4).
- The rule that takes you from one term to the next is called the term-to-term rule.

Test Yourself 1

Find the term-to-term rule and the next number in each of these sequences.

a
   4   7   10   13

b 15, 30, 45, 60

c 100, 90, 80, 70

d 2, 4, 8, 16

Solutions

Test Yourself 1

a The term-to-term rule is +3. The next term is 16.

b The term-to-term rule is +15. The next term is 75.

c The term-to-term rule is −10. The next term is 60.

d The term-to-term rule is ×2. The next term is 32.

Position-to-term rules

- You can use the formula for the nth term of a sequence to find any term without finding all the previous terms of the sequence.
- If you are given a formula for the nth term of a sequence and want to generate the sequence, you substitute 1 then 2 then 3 then 4 and so on into the formula to find the terms of the sequence.

Solution

Test Yourself 2

When $n = 1$, $5n − 2 = 3$.

When $n = 2$, $5n − 2 = 8$.

When $n = 3$, $5n − 2 = 13$.

When $n = 4$, $5n − 2 = 18$.

Test Yourself 2

The nth term of a sequence is $5n − 2$. Find the first four terms of the sequence.

Finding the *n*th term of a linear sequence

- The terms of a linear sequence increase or decrease by the same number each time.
- This number is called the common difference.
- Linear sequences which go up for example by 3 each time have an *n*th term of the form $3n + b$
- Linear sequences which go down for example by 2 each time have an *n*th term of the form $-2n + b$
- You can find *b* by substituting 1 for *n* and comparing the result with the first term.
- You can also work more formally by saying: The *n*th term of a linear sequence = Common difference $\times n$ + (First term − Common difference)
- You can write this as *n*th term = $An + b$
  - where *A* is the common difference
  - and *b* is the first term − the common difference

> **Chief Examiner says**
>
> Don't confuse the term-to-term rule with the rule for the *n*th term. A very common mistake is to say that the *n*th term of a sequence such as 2, 5, 8, 11, … is $n + 3$ instead of $3n - 1$.

Test Yourself 3

Find the *n*th term of these sequences.

a 7, 11, 15, 19, …

b 10, 7, 4, 1, −2 …

Use your answers to find the 50th terms.

Solutions

Test Yourself 3

a The terms increase by 4 each time so the *n*th term = $4n + b$.

Substituting $n = 1$ gives $4 + b$. The first term is 7 so $b = 3$.

So *n*th term = $4n + 3$.

50th term = $4 \times 50 + 3 = 203$

b The terms decrease by 3 each time so the *n*th term = $-3n + b$.

Substituting $n = 1$ gives $-3 + b$. The first term is 10 so $b = 13$.

So *n*th term = $-3n + 13$ or $13 - 3n$.

50th term = $13 - 3 \times 50 = -137$

Some special sequences

- Linear sequences include the even numbers, the odd numbers and multiples. Linear sequences increase or decrease by a constant amount, for example
  2, 5, 8, 11, 14, … or 6, 4, 2, 0, −2, …
- Square numbers: 1, 4, 9, 16, …
  The *n*th term of the square numbers is n^2.
- Powers of 2: 2, 4, 8, 16, …
- Powers of 10: 10, 100, 1000, 10 000, …
- Triangular numbers: 1, 3, 6, 10, 15, …

12 Constructions and loci

Constructions

Constructing triangles and quadrilaterals

- You need to be able to construct a triangle given one of four different sets of information.

  - Three sides

    Step 1: Draw line AB of given length.

    Step 2: Use compasses to construct arcs AC and BC with the compasses set to the given lengths.

    Step 3: Draw AC and BC.

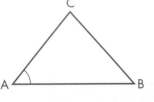

Arc with centre A.

Arc with centre B.

  - Two sides and the angle between them

    Step 1: Draw line AB of given length.

    Step 2: Measure the given angle at A.

    Step 3: Draw line AC of given length.

    Step 4: Join C to B.

  - Two angles and one side

    Step 1: Draw line AB of given length.

    Step 2: Measure the given angles at A and B.

    Step 3: Draw lines AC and BC.

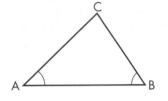

  - Two sides and an angle

    Step 1: Draw line AB of given length.

    Step 2: Measure the given angle at A.

    Step 3: Draw a line from A towards C.

    Step 4: To fix C, draw an arc at B, the radius of the arc being the second given length.

    This construction may give two possible triangles, ABC or ABC'.

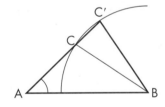

- You can use similar methods to construct quadrilaterals.

> **Chief Examiner says**
>
> You need to be able to draw and measure lines accurate to the nearest millimetre and angles to the nearest degree.

Test Yourself 1

Construct an isosceles triangle ABC with sides AB = 4 cm, BC = 5 cm and AC = 5 cm. Measure angle ABC of the triangle.

Test Yourself 2

Construct the quadrilateral ABCD where AB = 4.5 cm, angle BAD = 98°, angle ABC = 72°, BC = 3.4 cm and CD = 5.1 cm. Measure side AD of the quadrilateral.

> **Chief Examiner says**
>
> When you do constructions, remember to leave in all your construction lines. Do not rub them out.

Test Yourself 1

Step 1: Draw the 4 cm side of the triangle and label it AB.

Step 2: Set your compasses to 5 cm. With the compass point at A, make arcs above the centre of the line. Repeat with the point at B so that the arcs cross. This is point C.

Step 3: Draw AC and BC to complete the triangle.

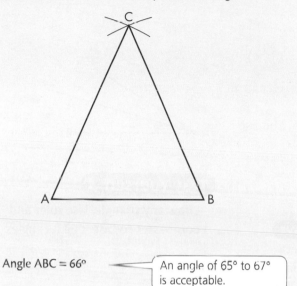

Angle ABC = 66° — An angle of 65° to 67° is acceptable.

Test Yourself 2

Step 1: Draw the 4.5 cm side of the quadrilateral and label it AB.

Step 2: Measure angles at A and B so that angle BAD = 98° angle ABC = 72°. Draw sides.

Step 3: Measure 3.4 cm from B and mark C on the side you have drawn.

Step 4: Set your compasses to 5.1 cm and with centre C draw an arc to cut the side you drew at A (you may need to extend the line). This is point D.

Step 5: Join sides AD and CD.

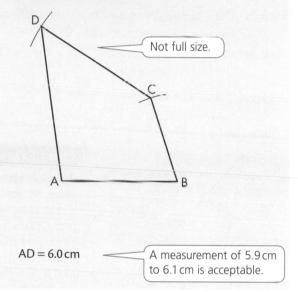

Not full size.

AD = 6.0 cm — A measurement of 5.9 cm to 6.1 cm is acceptable.

Other standard ruler and compasses constructions

You need to be able to do these constructions, using ruler and compasses.

- The perpendicular bisector of a line
  AB is the given line.
  Step 1: Draw two arcs of the same radius above and below the line, centred on A and B.
  Step 2: Join the intersections of the arcs. This line is the perpendicular bisector of the line AB.

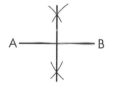

- The perpendicular from a point on a line
  AB is the given line. P is the given point.
  Step 1: Draw an arc, centred on P, to cut AB twice.
  Step 2: Open your compasses to a larger radius. Draw two arcs of the same radius, centred on the points of intersection of the first arc and AB, to cut each other above the line.
  Step 3: Join the point of intersection of the two arcs to P. This line is perpendicular to AB.

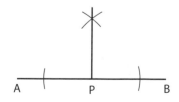

- The perpendicular from a point to a line
  AB is the given line. P is the given point.
  Step 1: Draw an arc, centred on P, to cut AB twice.
  Step 2: Draw two more arcs of the same radius, centred on the points of intersection of the first arc and AB, to cut each other below the line (on the opposite side from P).
  Step 3: Join the point of intersection of the two arcs to P. This line is perpendicular to AB.

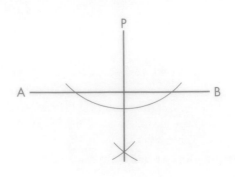

- The bisectors of an angle
  AB and AC are the two lines making the given angle.
  Step 1: Draw two arcs of the same radius, centred on A, to cut AB and AC.
  Step 2: Draw two more arcs of the same radius, centred on X and Y, to cut each other between AB and AC.
  Step 3: Join the point of intersection of the two arcs to A. This line is the bisector of angle A.

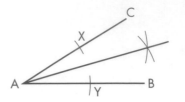

Solution

Test Yourself 3

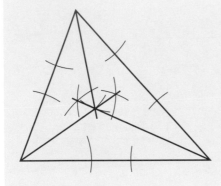

Chief Examiner says

If your construction is accurate, you should find that all the bisectors meet at one point. (This point is, in fact, the centre of a circle which touches each of the three sides of the triangle tangentially – you can check by drawing this circle, if you wish.)

Test Yourself 3

Draw any triangle. Use ruler and compasses to bisect all its angles.

Constructing a locus

A locus is a line or region where a point can be according to a rule.

- The locus of a point which is always 3 cm from a fixed point A is a circle, centre A, radius 3 cm.

- The locus of a point which is always less than 3 cm from a fixed point A is the region inside a circle, centre A, radius 3 cm.

- The locus of a point which is always more than 3 cm from a fixed point A is the region inside a circle, centre A, radius 3 cm.

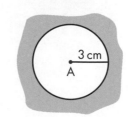

- The locus of a point which is an equal distance from two given points is the perpendicular bisector of the line joining the two points.

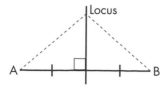

- The locus of a point which is an equal distance from two given intersecting lines is the bisectors of the angles between the lines.

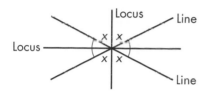

- The locus of a point which is 2 cm from a given straight line is the pair of lines parallel to that line.

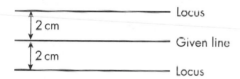

- The locus of a point which is 2 cm from a given line segment AB is the pair of parallel lines to that line segment joined by semicircles with centres at A and B and radius 2 cm.

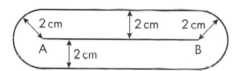

Test Yourself 4

2 cm

←——3 cm——→

Scale: 1 cm to 1 m

A goat is tethered by a 2 m rope to a rail 3 m long, which is fixed in a field of grass. The rope can slide along the rail. Draw a scale diagram to show the region of grass which the goat can eat.

Intersecting loci

Revised

- Exam questions often involve more than one locus.
- Draw one locus at a time.
- Show the final required region or points clearly by labelling, together with shading if necessary.

Find the locus of points equidistant from A, B and C.

(There is only one point on the page which is equidistant from A, B and C.)

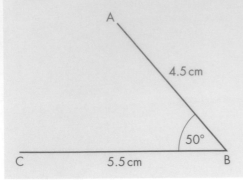

Test Yourself 5

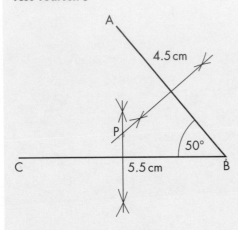

The locus of points equidistant from A and B is the perpendicular bisector of AB.

The locus of points equidistant from B and C is the perpendicular bisector of BC.

The required locus is where these two lines meet. It is marked P.

13 Sampling

Data sampling

- To investigate something about a population you may use a census or a sample.
- In a census you find out the information about every member of the population.
- In a sample, the information is obtained from a small proportion of the population, usually 10% or less. If the sample is too small, the results may not be reliable.
- To avoid bias, you need to ensure the sample is representative of the whole population.
- The advantage of a sample is that it is cheaper and quicker.
- The disadvantage is that, since it is not a census, its accuracy depends on how representative the sample is.
- There are various methods of sampling. Two of the most important are simple random sampling and stratified random sampling.

Simple random sampling

- For simple random sampling, every member of the population must have an equal chance of being chosen, irrespective of who has already been chosen.
- Give every member of a population a number and choose numbers at random.
- Random numbers can be chosen by a physical method, for example putting the numbers in a bag and choosing at random, or by using a random number generator on a calculator or computer.

Stratified random sampling

- In stratified random sampling, you set up strata that you want to ensure are fairly represented in your sample.
- For example, in a school you may want to ensure that every year group is represented fairly in the sample. So your strata will be the year groups.
- You then choose within those year groups using simple random sampling. So you may, for instance, choose at random 10% of each year group.

> **Chief Examiner says**
>
> The questions on this topic almost always ask you to describe a method or make comments. It is worth making sure you know the statements in the bullet points above.

Test Yourself 1

A company's workforce is divided into five divisions, as shown in the table.

Division	Design	Manufacture	Sales	Service	Maintenance
Number of workers	10	283	42	103	12

A stratified sample of 50 is to be selected. How many people should be selected from each division?

Solution

Test Yourself 1

Total workforce = 450

So a sample of 50 is a 1 in 9 sample.

Division	Design	Manufacture	Sales	Service	Maintenance
Number of workers	10	283	42	103	12
Number in sample	1	31	5	12	1

Bias

- If each member of the population does not have an equal chance of being selected for a sample then the sample is said to be biased.
- Bias can come from various sources.
  - The sample is not chosen randomly.
  - Only people who have time or people who feel strongly about the subject of the questionnaire may respond to it.
  - A replacement item is included. For example if you are interviewing every tenth household and there is nobody in you should not go to the next household where there is somebody in.
  - How, where or when data are collected may cause bias. For example, a survey done in a shoppers' car park excludes people who walk or get the bus to the shops.
  - When the questions are unclear or are leading questions.
- See Unit A Chapter 10 for more on designing a questionnaire.

Solution

Test Yourself 2

Everyone does not have an equal chance of being chosen. Some people are ex-directory, etc.

Test Yourself 2

Adam is carrying out a survey about use of public transport in his area. He decides to survey the first person on every page of the phone book.

Why is this **not** a simple random sample?

14 Trigonometry

- The three trigonometrical ratios are:

$$\sin x = \frac{\text{Opposite}}{\text{Hypotenuse}} \qquad \cos x = \frac{\text{Adjacent}}{\text{Hypotenuse}} \qquad \tan x = \frac{\text{Opposite}}{\text{Adjacent}}$$

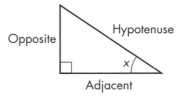

- When you are solving problems, you may need to use Pythagoras' theorem as well (see Unit A Chapter 7).

Test Yourself 1

a Find length x in this triangle.

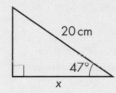

b Find angle y in this triangle.

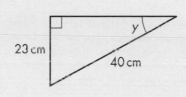

Chief Examiner says

- When using trigonometry, label the sides with H, O and A. This will help you work out which formula to use.
- When finding angles remember to use the $\sin^{-1}$, $\cos^{-1}$ and $\tan^{-1}$ keys. These are often found by pressing $\boxed{\text{SHIFT}}$, $\boxed{\text{INV}}$ or $\boxed{\text{2nd F}}$ followed by the $\boxed{\text{sin}}$, $\boxed{\text{cos}}$ or $\boxed{\text{tan}}$ keys.
- For accuracy, always use all the figures from one answer when you are using that value in another calculation.

Solutions

Test Yourself 1

a $\cos 47° = \dfrac{x}{20}$

$\qquad x = 20 \cos 47°$

$\qquad x = 13.6\,\text{cm}$

b $\sin y = \dfrac{23}{40}$

$\qquad y = \sin^{-1}\left(\dfrac{23}{40}\right)$

$\qquad y = 35.1°$

15 Representing and interpreting data

Cumulative frequency graphs

Revised

- Cumulative frequency is the running total of the frequencies in a distribution. The last cumulative frequency is the total of the frequencies and is often given in the question.
- Cumulative frequency is plotted at the upper bound of each interval. The points can be joined by a curve or by straight lines. The graph shows a fairly typical shape for a cumulative frequency curve.
- To find the median of a frequency distribution, draw a line across the graph at half the total frequency to meet the curve and then down to read off the value on the horizontal axis.
- To find the quartiles of a frequency distribution, draw lines across the graph at a quarter and three-quarters of the total frequency to meet the curve and then down to read off the values on the horizontal axis.
- Interquartile range (upper quartile – lower quartile) is a measure of spread.
- Box plots are a useful way of displaying data. They are also called 'box-and-whisker' plots.

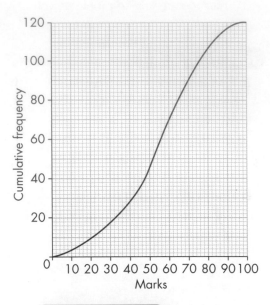

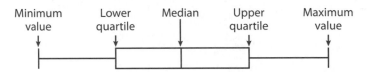

Comparing distributions

- To compare the two sets of data make two comparisons. State which set of data
  - is bigger on average using the median
  - is more spread out using the interquartile range.

Test Yourself 1

The table shows the distribution of the masses (mg) of 80 tomatoes.

a Calculate the cumulative frequencies.

b Draw a cumulative frequency diagram.

c Estimate the number of tomatoes weighing more than 93 g.

d Find the median and interquartile range.

e Draw a box plot.

Mass (mg)	Frequency
$70 < m \leqslant 75$	4
$75 < m \leqslant 80$	10
$80 < m \leqslant 85$	22
$85 < m \leqslant 90$	27
$90 < m \leqslant 95$	14
$95 < m \leqslant 100$	3

Test Yourself 1

a

Mass (*m*g)	Frequency	Cumulative frequency
$70 < m \leqslant 75$	4	4
$75 < m \leqslant 80$	10	14
$80 < m \leqslant 85$	22	36
$85 < m \leqslant 90$	27	63
$90 < m \leqslant 95$	14	77
$95 < m \leqslant 100$	3	80

b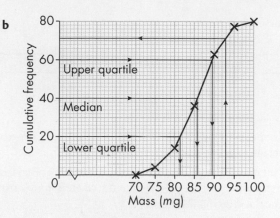

c $80 - 71 = 9$

Read off the frequency for 93 g from the graph. This gives you an estimate of the number of tomatoes with a mass of more than 93 g.

d Median = 86 g
Interquartile range = $89.5 - 81.5 = 8$ g

e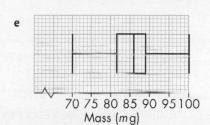

Test Yourself 2

This box plot summarises the times, in minutes, taken for 100 students at Fairlands College to get to school.

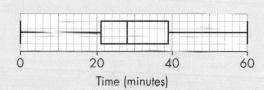

The table below summarises the times, in minutes, taken for the students at Breidon School to get to school.

Minimum	Lower quartile	Median	Upper quartile	Maximum
5	15	25	41	80

a Using the same scale as above, draw a box plot for Breidon School.

b Make two comparisons between the distribution of times for the two schools.

Solutions

Test Yourself 2

a

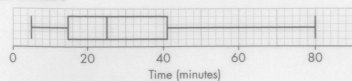

Time (minutes)

b Two valid comparisons such as:

1 The median shows that the students at Fairlands College took longer on average to get to school.

2 The times are more consistent for the students at Fairlands College as the interquartile range is only 18 minutes, whereas for the students at Breidon School it is 26 minutes.

Histograms

Revised ☐

- You use histograms when you have unequal intervals.
- A histogram is similar to a frequency diagram except that on a histogram the area of the bar is equal to the frequency.
- Width of interval × Height of bar = Frequency
- $\text{Height} = \dfrac{\text{Frequency}}{\text{Width of interval}}$
- The height is called the frequency density.

Solutions

Test Yourself 3

a

Time (t seconds)	Frequency (f)	Frequency density = f ÷ width
$0 < t \leqslant 2$	28	14
$2 < t \leqslant 5$	48	16
$5 < t \leqslant 10$	12	2.4
$10 < t \leqslant 15$	7	1.4
$15 < t \leqslant 20$	5	1

b

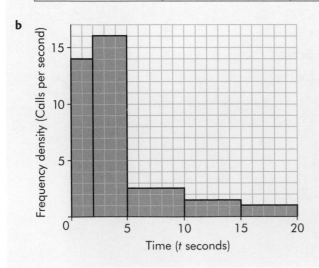

Test Yourself 3

The table shows the distribution of times (t seconds) that a company took to answer 100 telephone calls.

Time (t seconds)	Frequency (f)
$0 < t \leqslant 2$	28
$2 < t \leqslant 5$	48
$5 < t \leqslant 10$	12
$10 < t \leqslant 15$	7
$15 < t \leqslant 20$	5

a Calculate the frequency densities.

b Draw a histogram to represent the data.

- Look at the labels to tell you what the graph is about.
- Watch for exaggeration in the scales and scales that don't start at zero, which may affect your interpretation.

Test Yourself 4

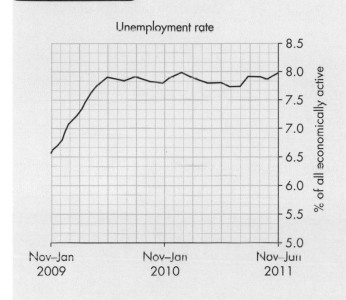

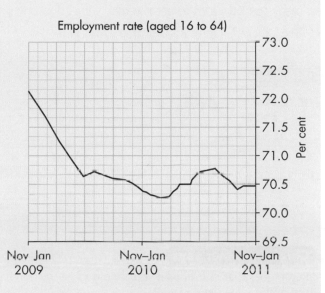

Source: Office for National Statistics

a Do the graphs show a difference in employment levels?

b Comment on the presentation.

Solutions

Test Yourself 4

a The trend is comparable but the rate of unemployment shown in the second graph is much higher (nearly 30% rather than 8%).

b The 'reversed' graphs could be confusing; the scales are the same making comparison easier; it is difficult to read the time scale; although the axes do not start at zero, it is justified as otherwise variation would not show and there is no deception; the population of the two graphs is different.

Chief Examiner says

You may think of other comments for part **b**). You would not be expected to have a complete list for full marks.

16 Formulae 2

Rearranging formulae

- To change the subject of a formula, use the equation rule of doing the same to both sides to get the new subject on one side of the formula.
- See Unit A Chapter 8 for simple examples.

Test Yourself 1

Make u the subject of this formula.

$$s = ut + \frac{1}{2}at^2$$

Solution

Test Yourself 1

$s = ut + \frac{1}{2}at^2$

$2s = at^2 + 2ut$ — Multiply both sides by 2, to eliminate the fractions.

$2s - at^2 = 2ut$ — Subtract at^2 from both sides.

$\dfrac{2s - at^2}{2t} = u$ — Divide both sides by $2t$. The fraction line acts as a bracket.

$u = \dfrac{2s - at^2}{2t}$ — Rewrite with u on the left.

Chief Examiner says

Remember that in formulae you should use a fraction line for divide not a divide ($\div$) sign.

Formulae where the new subject occurs more than once

- If the required subject is in the formula twice, carry out the following steps:
  - Rearrange so that all the terms involving the subject are on one side of the equation and all the other terms are on the other side.
  - Take the subject out as a common factor.
  - Divide both sides by the bracket.
  - If necessary, rewrite with the subject on the left.

Test Yourself 2

Make x the subject of $ax - by = cx + d$.

Solution

Test Yourself 2

$ax - by = cx + d$

$ax = cx + d + by$ — Add by to both sides.

$ax - cx = d + by$ — Subtract cx from both sides.

$x(a - c) = d + by$ — Take x out as a common factor.

$x = \dfrac{d + by}{a - c}$ — Divide both sides by $(a - c)$.

Formulae where the new subject is raised to a power

Revised ☐

- If the subject is raised to a power, for example v^2, first make v^2 the subject and then find the square root of both sides.
- For cubes find the cube root, for power four the fourth root and so on.
- If the subject is in a square root, for example $\sqrt{a}$, first make $\sqrt{a}$ the subject and then square both sides.

Test Yourself 3

Make u the subject of $v^2 = u^2 + 2as$.

Solution

Test Yourself 3

$v^2 = u^2 + 2as$

$v^2 - 2as = u^2$ —— (Subtract $2as$ from both sides.)

$\sqrt{v^2 - 2as} = u$ —— (Find the square root of both sides.)

$u = \sqrt{v^2 - 2as}$ —— (Rewrite with u on the left.)

Function notation

Revised ☐

- If y is a function of x then it can be written $y = f(x)$.
- $f(4)$ means the value of the function when $x = 4$
- For the function $f(x) = x + 3$,
    - solve $f(x) = 10$ means solve the equation $x + 3 = 10$
    - write an expression for $2f(x)$ means multiply $f(x)$ by 2, so for this example $2f(x) = 2(x + 3) = 2x + 6$
    - write an expression for $f(2x + 4)$ means use $(2x + 4)$ in place of x, so for this example $f(2x + 4) = (2x + 4) + 3 = 2x + 7$.

Test Yourself 4

$f(x) = 3x + 4$

a Find $f(5)$.

b Solve $f(x) = 10$.

c Write a simplified expression for $f(1 + x)$.

Solutions

Test Yourself 4

a $f(5) = 3 \times 5 + 4 = 19$

b $f(x) = 10$

    $3x + 4 = 10$

       $3x = 6$

        $x = 2$

c $f(1 + x) = 3(1 + x) + 4$

          $= 3 + 3x + 4$

          $= 3x + 7$

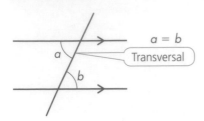

1 Properties of shapes

Angles made with parallel lines ⟶ Revised ☐

- *a* and *b* are called alternate angles. They are on opposite sides of the transversal.

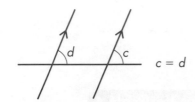

a = b · Transversal

- *c* and *d* are called corresponding angles. They are in the same position between the transversal and the parallel lines.

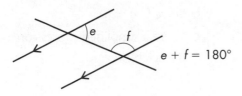

c = d

- *e* and *f* are called allied angles. They are the same side of the transversal between the parallel lines.

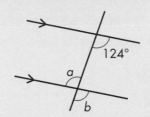

e + f = 180°

- You also need to know and be able to use these angle facts:
  - Angles around a point add up to 360°.
  - Vertically opposite angles are equal.

Solution

Test Yourself 1

a = 124° Alternate angles are equal.

b = 124° Corresponding angles are equal *or*
Vertically opposite angles are equal.

c = 44° Allied angles add up to 180°.

d = 84° Angles in a triangle add up to 180°.

e = 76° Corresponding angles are equal.

Test Yourself 1

Work out the angles marked with letters in these diagrams.

Give reasons for your answers.

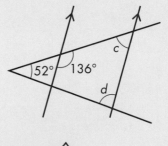

124°
a
b

52° 136°
c
d

63°
76° 41° *e*

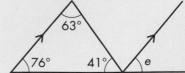

The angles in a triangle

Revised

- You need to know and be able to use these angle facts:
    - Angles on a straight line add up to 180°.
    - The angles in a triangle add up to 180°.
- The last two of these facts can be combined to show:
    - The exterior angle of any triangle is equal to the sum of the opposite interior angles.

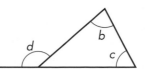

In the diagram, d is the exterior angle. So $d = b + c$.

Chief Examiner says

When you give reasons, make sure you give the relevant geometrical reason, not just a calculation. For example, write 'the angles in a triangle add up to 180°' not just '$x + 60 + 48 = 180$'.

Test Yourself 2

Work out the angles marked with letters in these diagrams.

Give reasons for your answers.

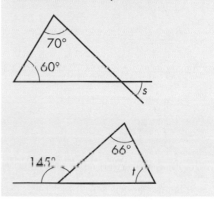

Solution

Test Yourself 2

$s = 50°$ Angles in a triangle add up to 180° *and* Vertically opposite angles are equal.

$t = 79°$ Exterior angle of triangle equals sum of interior opposite angles.
(Or angles on a straight line add up to 180° *and* angles in a triangle add up to 180°.)

The angles in a quadrilateral

Revised

- The sum of the interior angles of a quadrilateral is 360°.

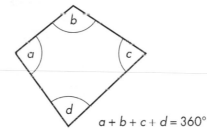

$$a + b + c + d = 360°$$

Test Yourself 3

Work out the size of angle a.

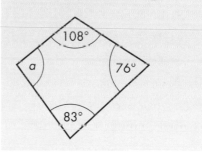

Solution

Test Yourself 3

$a = 360 - 108 - 76 - 83$

The angles of a quadrilateral add up to 360°.

$a = 93°$

Special quadrilaterals

These are the different quadrilaterals and the facts you need to know.

- Square
  - All angles 90°.
  - All sides equal.
  - Opposite sides parallel.
  - Diagonals equal and bisect at 90°.
  - Four lines of symmetry.
  - Rotation symmetry order 4.

- Rectangle
  - All angles 90°.
  - Opposite sides equal and parallel.
  - Diagonals equal and bisect but not at 90°.
  - Two lines of symmetry.
  - Rotation symmetry order 2.

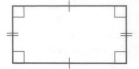

- Parallelogram
  - Opposite angles equal.
  - Opposite sides equal and parallel.
  - Diagonals not equal but bisect, though not at 90°.
  - No line symmetry.
  - Rotation symmetry order 2.

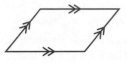

- Rhombus
  - Opposite angles equal.
  - All four sides equal.
  - Opposite sides parallel.
  - Diagonals not equal length but bisect at 90°.
  - Two lines of symmetry.
  - Rotation symmetry order 2.

- Trapezium
  - One pair of opposite sides parallel.

- Isosceles trapezium
  - Two pairs of adjacent angles equal.
  - One pair of opposite sides equal.
  - Other pair of sides parallel.
  - Diagonals equal but do not bisect or cross at 90°.
  - One line of symmetry.
  - No rotation symmetry.

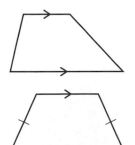

- Kite
  - One pair of opposite angles equal.
  - Two pairs of adjacent sides equal but none parallel.
  - One diagonal bisected at 90°.
  - One line of symmetry.
  - No rotation symmetry.

- Arrowhead
  - One reflex angle.
  - One pair of opposite angles equal.
  - Two pairs of adjacent sides equal but none parallel.
  - One diagonal bisected at 90°.
  - One line of symmetry.
  - No rotation symmetry.

Chief Examiner says

Remember what each shape is called and what it looks like. You can then use a sketch and symmetry to help you see which angles are equal, and so on.

Test Yourself 4

A quadrilateral has both pairs of opposite sides parallel.

What types of quadrilateral can it be?

Test Yourself 5

Draw two quadrilaterals which have only one line of symmetry.

Solutions

Test Yourself 4

It can be a square, a rectangle, a parallelogram or a rhombus.

Test Yourself 5

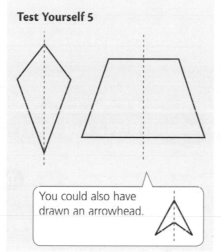

You could also have drawn an arrowhead.

The angles in a polygon

Revised

- The sum of the interior angles of a quadrilateral is 360°.
- The sum of the interior angles of a pentagon (five sides) is 540°.
- The sum of the interior angles of a hexagon (six sides) is 720°.
- At any vertex of a polygon, the interior and exterior angles make a straight line and add up to 180°.
- The sum of the exterior angles of any polygon is 360°.
- A regular polygon has all its sides equal and all its angles equal.
- The angle at the centre of a regular polygon = 360 ÷ Number of sides.

Test Yourself 6

A regular polygon has 12 sides.

a Work out the size of each exterior angle.

b Work out the size of each interior angle.

c Find the sum of the interior angles.

Solutions

Test Yourself 6

a Each exterior angle $= \frac{360}{12}$
$= 30°$

b Each interior angle $= 180° - 30°$
$= 150°$

c Sum of the interior angles $= 150° \times 12$
$= 1800°$

2 Fractions, decimals and percentages

Comparing fractions Revised

- Change the fractions to equivalent fractions with the same denominator.
- The fraction with the smaller numerator is the smaller fraction.

Solutions

Test Yourself 1

a $\frac{3}{5} = \frac{9}{15}$ and $\frac{2}{3} = \frac{10}{15}$

9 is less than 10 so $\frac{3}{5}$ is smaller than $\frac{2}{3}$.

b $\frac{7}{15} = \frac{28}{60}$ and $\frac{9}{20} = \frac{27}{60}$

27 is less than 28 so $\frac{9}{20}$ is smaller than $\frac{7}{15}$.

Test Yourself 1

Which is the smaller fraction in each pair?

a $\frac{3}{5}$ $\frac{2}{3}$

b $\frac{7}{15}$ $\frac{9}{20}$

Adding and subtracting fractions and mixed numbers Revised

- To add or subtract fractions,
  - change the fractions to equivalent fractions with the same denominator
  - then add or subtract the numerators.
- When mixed numbers are involved, deal with the whole numbers first.
- In a subtraction, if the first fraction is smaller than the second, change a whole number to a fraction.

Test Yourself 2

Work out these.

a $\frac{1}{6} + \frac{3}{4}$ **b** $\frac{4}{5} - \frac{2}{3}$

c $2\frac{2}{3} + 1\frac{3}{8}$ **d** $3\frac{1}{4} - 1\frac{2}{5}$

Solutions

Test Yourself 2

a $\frac{1}{6} + \frac{3}{4} = \frac{2}{12} + \frac{9}{12}$ ⟵ Change each fraction to an equivalent fraction with a denominator of 12.

$= \frac{11}{12}$ ⟵ Then add the numerators.

b $\frac{4}{5} - \frac{2}{3} = \frac{12}{15} - \frac{10}{15}$ ⟵ Change each fraction to an equivalent fraction with a denominator of 15.

$= \frac{2}{15}$ ⟵ Then subtract the numerators.

c $2\frac{2}{3} + 1\frac{3}{8} = 3 + \frac{2}{3} + \frac{3}{8}$ ⟵ First add the whole numbers.

$= 3 + \frac{16}{24} + \frac{9}{24}$ ⟵ Change each fraction to an equivalent fraction with a denominator of 24.

$= 3 + \frac{25}{24}$ ⟵ Add the numerators.

$= 4\frac{1}{24}$ ⟵ Change the improper fraction to a mixed number and add the whole numbers.

d $3\frac{1}{4} - 1\frac{2}{5} = 2 + \frac{1}{4} - \frac{2}{5}$ ⟵ Subtract the whole numbers.

$= 2 + \frac{5}{20} - \frac{8}{20}$ ⟵ Change each fraction to an equivalent fraction with a denominator of 20.

$= 1 + \frac{20}{20} + \frac{5}{20} - \frac{8}{20}$ ⟵ Change one of the whole numbers to a fraction.

$= 1\frac{17}{20}$ ⟵ Collect the fractions.

Multiplying and dividing fractions and mixed numbers

Multiplying fractions

- When multiplying a fraction by a whole number, multiply the numerator by the whole number and then simplify.
- When multiplying fractions, multiply the numerators, multiply the denominators and then simplify.

Chief Examiner says

Remember, $\frac{1}{2} \times \frac{1}{3} = \frac{1}{6}$. A common error is to multiply 1×1 and get 2.

Solutions

Test Yourself 3

a $\frac{5}{6} \times 9 = \frac{45}{6}$ or $\frac{5}{6} \times 9 = \frac{5}{2} \times 3$

$\qquad = \frac{15}{2} \qquad\qquad\qquad\quad = \frac{15}{2}$

$\qquad = 7\frac{1}{2} \qquad\qquad\qquad\quad = 7\frac{1}{2}$

> You can simplify either before or after multiplying.

b $\frac{4}{5} \times \frac{5}{9} = \frac{20}{45}$ or $\frac{4}{5} \times \frac{5}{9} = \frac{4}{1} \times \frac{1}{9}$

$\qquad = \frac{4}{9} \qquad\qquad\qquad\qquad = \frac{4}{9}$

Dividing fractions

- When dividing fractions, turn the second fraction upside down and multiply.
- Turning a fraction upside down forms the reciprocal of the fraction.

Solutions

Test Yourself 4

a $\frac{3}{4} \div 6 = \frac{3}{4} \div \frac{6}{1}$ ⟵ $6 = \frac{6}{1}$

$\qquad = \frac{3}{4} \times \frac{1}{6}$ ⟵ Turn $\frac{6}{1}$ upside down and mulitply.

$\qquad = \frac{1}{4} \times \frac{1}{2}$ ⟵ Cancel.

$\qquad = \frac{1}{8}$

b $\frac{3}{5} \div \frac{7}{10} = \frac{3}{5} \times \frac{10}{7}$ ⟵ Turn $\frac{7}{10}$ upside down and multiply.

$\qquad = \frac{3}{1} \times \frac{2}{7}$ ⟵ Cancel.

$\qquad = \frac{6}{7}$

Test Yourself 3

Work out these, giving the answers as simply as possible.

a $\frac{5}{6} \times 9$

b $\frac{4}{5} \times \frac{5}{9}$

Test Yourself 4

Work out these, giving the answers as simply as possible.

a $\frac{3}{4} \div 6$

b $\frac{3}{5} \div \frac{7}{10}$

Multiplying and dividing mixed numbers

- When mixed numbers are involved, first change to improper fractions.

Solutions

Test Yourself 5

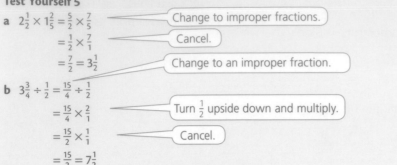

a $2\frac{1}{2} \times 1\frac{2}{5} = \frac{5}{2} \times \frac{7}{5}$ — Change to improper fractions.

$\quad = \frac{1}{2} \times \frac{7}{1}$ — Cancel.

$\quad = \frac{7}{2} = 3\frac{1}{2}$ — Change to an improper fraction.

b $3\frac{3}{4} \div \frac{1}{2} = \frac{15}{4} \div \frac{1}{2}$

$\quad = \frac{15}{4} \times \frac{2}{1}$ — Turn $\frac{1}{2}$ upside down and multiply.

$\quad = \frac{15}{2} \times \frac{1}{1}$ — Cancel.

$\quad = \frac{15}{2} = 7\frac{1}{2}$

Test Yourself 5

Work out these.

a $2\frac{1}{2} \times 1\frac{2}{5}$

b $3\frac{3}{4} \div \frac{1}{2}$

Adding and subtracting decimals

Revised

- Keep tenths under tenths, units under units, tens under tens, and so on.
- Make sure the decimal points are underneath each other.
- Always start from the column on the right.

Chief Examiner says

When a question says 'You must show your working', you will not get a mark if you just write down the answer.

Test Yourself 6

Work out these.

Show your working.

a $537.5 + 27.68$

b $53.78 - 24.9$

Solutions

Test Yourself 6

a $\quad 537.50$
$\quad + 27.68$
$\quad \overline{565.18}$

b $\quad 53.78$
$\quad - 24.90$ — Add a zero so that there is the same number of decimal places in each number.
$\quad \overline{28.88}$

Multiplying and dividing decimals

Revised

Multiplication

- You need to know the basic multiplication tables.
- To multiply numbers, you may have learnt different methods: long multiplication, multiplying by hundreds, tens and units separately and then adding up; using a grid; using a lattice. Make sure you know which method works for you.
- When multiplying a decimal number by 10, move the point one place to the right; when multiplying by 100, move the point two places to the right; and so on.

- When multiplying a decimal by a whole number or a decimal,
  - multiply ignoring the decimal points
  - put the same number of decimal places in the answer as in the question.

Test Yourself 7

Work out these.

a 2.714×100

b 5.67×2.3

Solutions

Test Yourself 7

a $2.714 \times 100 = 271.4$

b First multiply ignoring the decimal points.

Long multiplication

$$
\begin{array}{r}
567 \\
\times\, 23 \\
\hline
1\,701 \\
11\,340 \\
\hline
13\,041
\end{array}
$$

Multiply by 3.

Multiply by 20.

Add 1 701 and 11 340.

Using a grid

×	500	60	7
20	10 000	1 200	140
3	1 500	180	21

$10\,000 + 1\,200 + 140 + 1\,500 + 180 + 21 = 13\,041$

Using a lattice

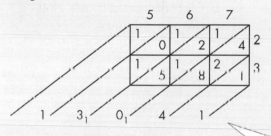

$5.67 \times 2.3 = 13.041$

Add along the diagonals, carrying if the total is greater than 10.

There are $2 + 1 = 3$ decimal places in the question so there must be 3 decimal places in the answer.

Division

- When dividing a decimal by 10, move the point one place to the left; when dividing by 100, move the point two places to the left; and so on.
- When dividing a decimal by a whole number, the decimal point in the answer goes above the decimal point in the question.
- When dividing a number by a decimal,
  - first multiply both numbers by 10 or 100 so that the number you are dividing by is a whole number
  - then divide.

Put the decimal points underneath each other if there is still a decimal point in the number you are dividing.

Test Yourself 8

Work out these.

a $8.17 \div 10$

b $46.8 \div 9$

c $14.56 \div 5.6$

Solutions

Test Yourself 8

a 0.817

b Set out your calculation with the decimal points underneath each other.

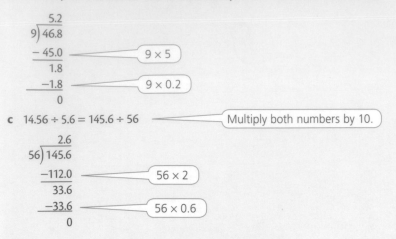

c 14.56 ÷ 5.6 = 145.6 ÷ 56 ————（Multiply both numbers by 10.）

```
      2.6
56) 145.6
   −112.0 ————（56 × 2）
     33.6
    −33.6 ————（56 × 0.6）
        0
```

Percentage increase and decrease   Revised

- To find a percentage of an amount, change the percentage to a decimal and multiply.

- To increase or decrease by a percentage, add or subtract the percentage to or from 100. Change this to a decimal (this is the multiplier) and multiply. Alternatively, work out the increase or decrease and add it on or subtract it.

Test Yourself 9

a £84 is increased by 7%. Find the new amount.

b £12 is decreased by 8%. Find the new amount.

Solutions

Test Yourself 9

a 100 + 7 = 107   or   7% as a decimal is 0.07
   Decimal = 1.07   0.07 × 84 = 5.88
   New amount   New amount
       = 84 × 1.07   = 84 + 5.88
       = £89.88   = £89.88

b 100 − 8 = 92   or   8% as a decimal is 0.08
   Decimal = 0.92   0.08 × 12 = 0.96
   New amount   New amount
       = 12 × 0.92   = 12 − 0.96
       = £11.04   = £11.04

3 Mental methods

Rounding to a given number of significant figures
Revised ☐

- In any number, the first digit from the left which is not 0 is the first significant figure.
- When rounding to a given number of significant figures,
  - if the first digit not required is 4 or less, then just ignore it and, if necessary, include the correct number of zeros to show the size of the number. For example, 0.3719 to 2 significant figures is 0.37 and 3719 to 2 significant figures is 3700.
  - if the first digit not required is 5 or more, add 1 to the last significant figure required and, if necessary, include the correct number of zeros to show the size of the number. For example, 0.062 85 to 2 significant figures is 0.063 and 6285 to 2 significant figures is 6300.

> **Chief Examiner says**
>
> The number 0.997 correct to 2 significant figures is 1.0 but correct to 1 significant figure it is 1. Extra zeros are used only to show the size of a number, as in 200, for example.

- You can use rounding to 1 significant figure to estimate answers to problems. Round each number to 1 significant figure and then carry out the calculation without a calculator.

> **Test Yourself 1**
>
> Write the following correct to 2 significant figures.
>
> **a** 3.658
>
> **b** 0.0543
>
> **c** 935.4
>
> **d** 427 510

> **Test Yourself 2**
>
> Estimate the answer to each of these calculations.
>
> **a** 47.5×12.6
>
> **b** $628 \div 43$
>
> **c** $\dfrac{24 \times 6.99}{3.4}$

Solutions

Test Yourself 1

a 3.7

b 0.054

c 940

d 430 000

Test Yourself 2

a $50 \times 10 = 500$

b $600 \div 40 = 15$

c $\dfrac{20 \times 7}{3} = \dfrac{140}{3} = 46.666 \ldots$

$\approx 50 \text{ or } 47$

Deriving unknown facts from those you know
Revised ☐

- When the solution to a calculation is known, the result of a similar or inverse calculation with the numbers multiplied or divided by powers of 10 can be found by multiplying or dividing by an appropriate power of ten.

Solutions

Test Yourself 3

a $47 \times 36 = 1692$

$\times 10, \times 10 = \times 100$

b $470 \times 3.6 = 1692$

$\times 100, \text{ no change} = \times 100$

c $1692 \div 4.7 = 360$

$\times 100, \text{ no change} = \times 100$

> **Test Yourself 3**
>
> Given that $4.7 \times 3.6 = 16.92$, find the value of each of these calculations.
>
> **a** 47×36
>
> **b** 470×3.6
>
> **c** $1692 \div 4.7$

4 Transformations

Chief Examiner says

At GCSE, you will meet four types of transformations: reflection, rotation, translation and enlargement.

You may be asked to follow instructions and draw the result of a transformation or to describe a transformation that has taken place. When describing a transformation, give the type of transformation first, then the extra information required.

Reflections

Revised

- The image is the same shape and size as the object, but is reversed.
- The distance between each point on the image and the mirror line is the same as the distance between the mirror line and the corresponding point on the object.
- When describing a reflection, state the mirror line.
- An object and its reflected image together form a shape with reflection symmetry. In two dimensions, the mirror is the line of symmetry. In three dimensions, the mirror is the plane of symmetry.

Test Yourself 1

Reflect this triangle in the line $x = 3$.

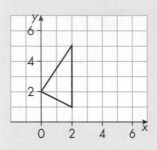

Solution

Test Yourself 1

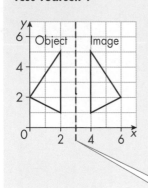

The line $x = 3$ is the vertical line through 3 on the x-axis.

Rotations

Revised

- The image is the same shape and size as the object, but is turned round.
- The distance between each point on the image and the centre of rotation is the same as the distance between the centre of rotation and the corresponding point on the object.
- Use tracing paper and trial and improvement to help you find the centre of rotation.
- When describing a rotation, state
  - the centre of rotation
  - the angle of rotation
  - the direction – clockwise or anticlockwise.

Test Yourself 2

Rotate this flag through 90° clockwise about (0, 2).

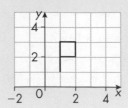

Test Yourself 2

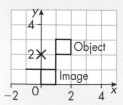

Translations

- Each point on the image has moved the same distance, in the same direction, from the corresponding point on the object.
- The image is the same size and shape as the object and is the same way up.
- When describing a translation, state the column vector, or how many units the shape has moved in each direction.

Solution

Test Yourself 3

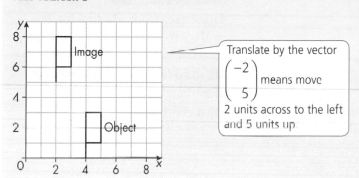

Translate by the vector $\begin{pmatrix} -2 \\ 5 \end{pmatrix}$ means move 2 units across to the left and 5 units up

Test Yourself 3

Translate this flag by $\begin{pmatrix} -2 \\ 5 \end{pmatrix}$

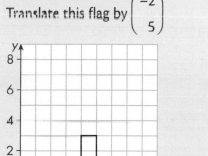

Enlargements

- The image is the same shape as the object but each length on the image is the corresponding length on the object multiplied by the scale factor.
- This is the only transformation you meet at GCSE where the object and the image are not congruent. When two things are the same shape but a different size like this, they are called similar shapes.
- The distance of each point on the object from the centre of enlargement is multiplied by the scale factor to find the distance from the centre to the corresponding points on the image.
- To find the centre of enlargement, join corresponding points on the object and image and extend the lines until they meet.
- When describing an enlargement, state
  - the centre of enlargement
  - the scale factor.

- If the scale factor is a fraction, the image is smaller than the object. It is still called an enlargement.
- A enlargement with a scale factor of $\frac{1}{3}$ is the inverse of an enlargement with a scale factor of 3.
- If the scale factor is negative, each point is transformed to a point on the other side of the centre of enlargement.

Solution

Test Yourself 4

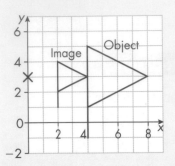

Solution

Test Yourself 5

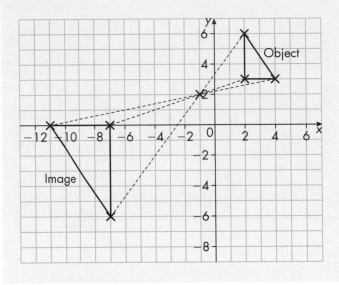

Test Yourself 4

Enlarge this flag with centre (0, 3) and scale factor $\frac{1}{2}$.

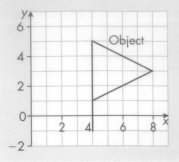

Test Yourself 5

Draw axes from −12 to 6 for x and −8 to 6 for y. Plot the points (2, 3), (4, 3) and (2, 6) and join them to form a triangle. Enlarge the triangle by scale factor −2 with centre (−1, 2).

Combining transformations

Revised ☐

- Perform the transformations in the order given.
- Compare the object and the final image to find the equivalent single transformation.

Chief Examiner says

In an examination, when asked to describe a single transformation, you have first to work out which type of transformation is needed. If the object and image are not the same size, check that it is an enlargement. If they are the same size, draw the object on tracing paper and move the tracing paper until it fits on the image. If you just move the tracing paper in a straight line, it is a translation. If you need to turn the tracing paper round, it is a rotation. If you turn the tracing paper over, it is a reflection.

Don't forget to give all the information needed to define the transformation.

Solution

Test Yourself 6

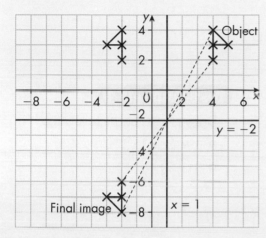

The transformation is a rotation of 180° about a centre (1, −2).

Draw axes from −8 to 6 for x and −8 to 4 for y. Plot the points (4, 2), (4, 3), (4, 4) and (5, 3) and join them to form a flag. Reflect the flag in the line $x = 1$. Then reflect the image in the line $y = −2$. Find the single transformation that is equivalent to these two reflections.

5 Straight-line graphs

Revised

Drawing straight-line graphs and Harder straight-line graphs

- Equations of straight lines are usually given in one of two forms: $y = 2x + 1$ or $3x + 2y = 6$.
- As with all graphs, the first step is to make a table of the x values and y values.
- With the $y = 2x + 1$ type, choose about three simple values of x.
- With the $3x + 2y = 6$ type, chose $x = 0$ and $y = 0$. You may wish to choose another value of x as a check.
- Next plot the coordinate pairs.
- Finally join the points up with a straight line and extend it to cover the grid.

Test Yourself 1

Draw graphs of the following.

a $y = 2x + 1$

b $3x + 2y = 6$

Solutions

Test Yourself 1

a

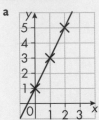

$x = 2, y = 2 \times 2 + 1 = 5$

x	0	1	2
y	1	3	5

b

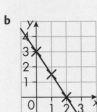

For $x = 0$, $2y = 6$ so $y = 3$

x	0	1	2
y	3	$1\frac{1}{2}$	0

For $y = 0$, $3x = 6$ so $x = 2$

For $x = 1$, $3 + 2y = 6$ so $y = 1\frac{1}{2}$

Distance–time graphs

- The steeper the line, the greater the speed.
- You can find speed from the relationship: Speed = Distance ÷ Time.
- A horizontal line indicates that the object is not moving.

Test Yourself 2

Asif walked to the bus stop, waited for the bus, then travelled on to school. The graph is a distance–time graph for Asif's journey.

a How long did Asif wait at the bus stop?

b How far did Asif travel

   i on foot?   **ii** by bus?

c How fast did Asif walk?

d What was the average speed of the bus?

e Why are the sections of the graph, in reality, unlikely to be straight?

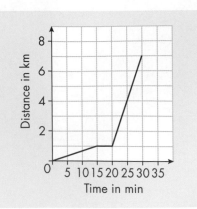

Solutions

Test Yourself 2

a 5 minutes

b **i** 1 km   **ii** 6 km

c $\frac{1}{15}$ km/min or 4 km/h

d $\frac{6}{10}$ km/min or 36 km/h

e Stops and starts are likely to be gradual, not sudden.
Speed is unlikely to be constant.

> **Chief Examiner says**
>
> Remember to find the rate of change for each part of the graph. If it is zero, say how long this lasts.

Finding the gradient of a straight-line graph · · · · · · · Revised ☐

* The gradient of a line is a number indicating how steep it is.
  The larger the number the steeper the line.
* Lines with a positive gradient slope forwards.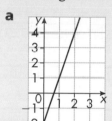
* Lines with a negative gradient slope backwards.
* Gradient $= \dfrac{\text{increase in } y}{\text{increase in } x}$.

Test Yourself 3

Find the gradient of these lines.

a

b

Solutions

Test Yourself 3

a Gradient $= \frac{6}{2} = 3$ **b** Gradient $= \frac{-4}{2} = -2$

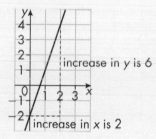
increase in y is 6
increase in x is 2

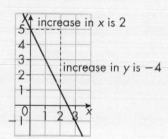
increase in x is 2
increase in y is −4

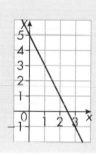

Finding the equation of a straight-line graph · · · · · · · Revised ☐

* The y-intercept is the value of y where the line crosses the y-axis.
* The general equation of a straight line is $y = mx + c$.
* In the equation, m stands for the gradient of the line and c is the y-intercept.
* The equation of a line **must** be written in the form $y = mx + c$ for the two numbers to represent the gradient and the y-intercept.

Test Yourself 4

a i Gradient = 3, y-intercept = -2
Equation is $y = 3x - 2$

ii Gradient = -2, y-intercept = 5
Equation is $y = -2x + 5$

b i $y = x + 4$
$m = 1$, $c = 4$

ii $2y = 6x - 3$
$y = 3x - 1\frac{1}{2}$ ⟵ Divide both sides by 2.
$m = 3$, $c = -1\frac{1}{2}$

iii $5x - 2y = 12$
$5x - 12 = 2y$ ⟵ Rearrange to make y the subject.
$y = 2\frac{1}{2}x - 6$
$m = 2\frac{1}{2}$, $c = -6$

Test Yourself 4

a Find the equations of these lines.

i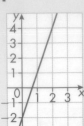

ii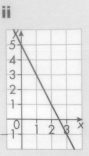

b Find the gradient (m) and y-intercept (c) of these lines.

i $y = x + 4$
ii $2y = 6x - 3$
iii $5x - 2y = 12$

Exploring gradients

Revised

- Lines which are parallel have the same gradient.
- If two lines with gradients m_1 and m_2 are perpendicular then $m_1 m_2 = -1$.
- If a line has gradient m, then a line perpendicular to it has gradient $-\dfrac{1}{m}$.

Test Yourself 5

a From $y = 2x - 5$, $\quad m = 2$
From $(0, 4)$, $\quad c = 4$
Equation is $y = 2x + 4$

b $m_1 = 4 \quad$ so $m_2 = -\frac{1}{4}$.

Equation is $y = -\frac{1}{4}x + c$

Since line goes through $(8, 5)$, substituting
gives $5 = \left(-\frac{1}{4}\right) \times 8 + c$
$\quad 5 = -2 + c$ so $c = 7$
So equation is $y = -\frac{1}{4}x + 7$ or $4y + x = 28$

Test Yourself 5

a Find the equation of the line parallel to $y = 2x - 5$ which passes through the point $(0, 4)$.

b Find the equation of the line perpendicular to $y = 4x + 1$ which passes through the point $(8, 5)$.

6 Indices, decimals and surds

The rules of indices
Revised

- The rules of indices are
  - $a^m \times a^n = a^{m+n}$
  - $a^m \div a^n = a^{m-n}$
  - $(a^m)^n = a^{m \times n}$
  - $a^0 = 1$
  - $a^{-n} = \dfrac{1}{a^n}$
  - $a^{\frac{1}{n}} = \sqrt[n]{a}$
  - $a^{\frac{m}{n}} = \sqrt[n]{a^m} = (\sqrt[n]{a})^m$
- Remember $a^1 = a$.

Chief Examiner says

For adding and subtracting, indices are not so helpful. There is no shortcut rule. So, for example $3^3 + 3^2 = 27 + 9 = 36$.

Test Yourself 1

Simplify the following.

a $3^2 \times 3^4$

b $\dfrac{3^5}{3^3}$

c $(2^3)^4$

Test Yourself 2

Evaluate the following.

a 16^{-1}

b $27^{\frac{1}{3}}$

c $16^{\frac{3}{2}}$

Solutions

Test Yourself 1

a 3^6 — Add the powers: $2 + 4 = 6$

b 3^2 — Subtract the powers: $5 - 3 = 2$

c 2^{12} — Multiply the powers: $3 \times 4 = 12$

Test Yourself 2

a $16^{-1} = \dfrac{1}{16^1} = \dfrac{1}{16}$

b $27^{\frac{1}{3}} = \sqrt[3]{27} = 3$

c $16^{\frac{3}{2}} = (\sqrt{16})^3 = 4^3 = 64$

Using the rules of indices with numbers and letters
Revised

- You need to be able to use the rules of indices with numbers and letters.

Solutions

Test Yourself 3

a x^7 — Add the powers.

b x^3 — Subtract the powers.

c 1

d x^6 — Multiply the powers.

e $10x^5$ — $2 \times 5 = 10$. Add the powers.

f x^6 — Add the powers to give x^8 on the top, then $8 - 2 = 6$

Test Yourself 3

Simplify the following.

a $x^5 \times x^2$

b $x^6 \div x^3$

c x^0

d $(x^2)^3$

e $2x^2 \times 5x^3$

f $\dfrac{x^3 \times x^5}{x^2}$

Terminating and recurring decimals
Revised

- All fractions are equal to either recurring or terminating decimals.
- The fractions equal to terminating decimals have denominators whose only prime factors are 2 and/or 5.

- To convert a fraction to a decimal, divide the numerator by the denominator.
- You can use dot notation to write recurring decimals. For example, $\frac{1}{3} = 0.\dot{3}$.
- If more than one digit recurs, you put a dot over the first and last digit of the group, or period. For example, $0.012\,612\,6 = 0.0\dot{1}2\dot{6}$.

Test Yourself 4

Write these fractions as decimals.

a $\frac{1}{8}$ b $\frac{1}{6}$

c $\frac{13}{20}$ d $\frac{7}{25}$

e $\frac{3}{7}$

Solutions

Test Yourself 4

a 0.125 — Terminating, prime factor 2.

b 0.1$\dot{6}$ — Recurring, prime factors 2 and 3.

c 0.65 — Terminating, prime factors 2 and 5.

d 0.28 — Terminating, prime factor 5.

e 0.$\dot{4}$28 57$\dot{1}$ — Recurring, prime factor 7.

Changing a recurring decimal to a fraction

- The method for doing this is best illustrated by an example.
  Express $0.\dot{4}\dot{2}$ as a fraction in its lowest terms.

  Let $r = 0.\dot{4}\dot{2}$

  So
  $$r = 0.424\,242\,42...$$
  $$100r = 42.424\,242\,42...$$

  Subtracting $99r = 42$

  $$r = \frac{42}{99} = \frac{14}{33}$$

Chief Examiner says

Multiply by 10^n, where n is the number of recurring figures.

Test Yourself 5

Express $0.\dot{5}0\dot{7}$ as a fraction in its lowest terms.

Solution

Test Yourself 5

Let $r = 0.\dot{5}0\dot{7}$

So
$$r = 0.507\,507\,507...$$
$$1000r = 507.507\,507\,507...$$

Subtracting $999r = 507$

$$r = \frac{507}{999} = \frac{169}{333}$$

Surds

Revised

- A surd is an expression involving a square root sign.
  For example $\sqrt{3}$ or $5 + \sqrt{2}$
- An irrational number is a decimal that neither terminates nor recurs.
  For example, $\sqrt{2}$ is an irrational number. If you are asked to give an exact answer, leave your answer in surd form.
- When adding or subtracting surds, deal with the rational parts and the irrational parts separately.
- When simplifying surds, look for the highest factor that has an exact square root.
- When multiplying surds, use the normal rules of algebra.
- To rationalise the denominator of a fraction with a simple surd in the denominator, multiply the numerator and denominator by the surd.

Test Yourself 6

Simplify the following.

a $\sqrt{80}$

b $\sqrt{3}\,(5 + 2\sqrt{3})$

c $\dfrac{6}{\sqrt{15}}$

Test Yourself 6

a $\sqrt{80} = \sqrt{16} \times \sqrt{5}$

$\quad = 4\sqrt{5}$ ——— Use the result $\sqrt{ab} = \sqrt{a} \times \sqrt{b}$

b $\sqrt{3}\,(5 + 2\sqrt{3}) = 5\sqrt{3} + 2 \times \sqrt{3} \times \sqrt{3}$

$\qquad\qquad\quad = 5\sqrt{3} + 6$ ——— $2 \times \sqrt{3} \times \sqrt{3} = 6$

c $\dfrac{6}{\sqrt{15}} = \dfrac{6 \times \sqrt{15}}{\sqrt{15} \times \sqrt{15}} = \dfrac{6\sqrt{15}}{15} = \dfrac{2\sqrt{15}}{5}$

To simplify this expression you have to rationalise the denominator.

7 Inequalities

Solving inequalities with one unknown

- These are the inequality symbols:
  - $>$ means 'is greater than'
  - $<$ means 'is less than'
  - $\geqslant$ means 'is greater than or equal to'
  - $\leqslant$ means 'is less than or equal to'
- Inequalities are generally solved using the same rules as for solving equations.
- The exception is when multiplying and dividing by a negative number. In this case the inequality sign is reversed. This problem is best avoided. You do this by moving the unknown to the side where it is positive.
- Inequalities can be shown on a number line using these conventions:
  - $\circ\!\!\longrightarrow$ when 'equals' is not included
  - $\bullet\!\!\longrightarrow$ when 'equals' is included

Chief Examiner says

Don't forget that the solution to an inequality is itself an inequality such as $x \geqslant 4$.

You will lose marks if you simply give a value of x.

Test Yourself 1

a List the integers for which $-2 \leqslant x < 3$.

b Solve the following.

 i $3x - 1 \geqslant 11$

 ii $5 - x < 2$

 iii $x + 2 \geqslant 3 - x$

Show your solutions on a number line.

Solutions

Test Yourself 1

a $-2, -1, 0, 1, 2$ — 2 is included but 3 is not.

b i $3x - 1 \geqslant 11$
  $3x \geqslant 11 + 1$
  $3x \geqslant 12$
  $x \geqslant 4$

'Equals' included so the circle is filled in.

ii $5 - x < 2$
  $5 < 2 + x$ — Add x to both sides so that x is positive.
  $3 < x$ — Subtract 2 from both sides.
  $x > 3$ — Swap sides so x is on the left.

'Equals' not included so use an open circle.

iii $x + 2 \geqslant 3 - x$
  $2x + 2 \geqslant 3$
  $2x \geqslant 1$
  $x \geqslant 0.5$

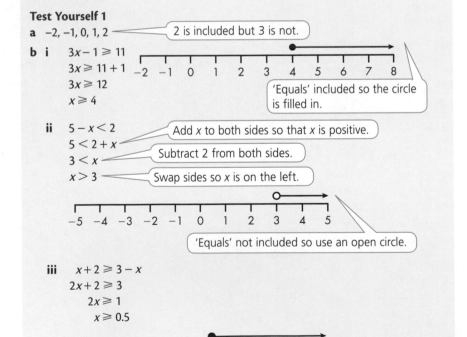

Solving inequalities with two unknowns

- You can show inequalities with two variables on a graph.

  Step 1: Draw the boundary lines. The equations are the inequalities with the inequality signs replaced with equals signs.

  If the inequality sign is $\leqslant$ or $\geqslant$, use a solid line.

  If the inequality sign is $<$ or $>$, use a dotted line.

  Step2: Decide which side of each line represents the given inequality.

  If the line does not go through the point $(0, 0)$, this is the best point to use as a test.

  If the line does go through the point $(0, 0)$, then use another point such as $(1, 0)$.

  Step 3: Indicate clearly the single region which satisfies all of the inequalities. Shade the unwanted side of the line. The part of the graph left unshaded is the required region.

Chief Examiner says

Make sure you show the required region clearly. If you are using the 'shaded out' convention, state that this is the case.

Test Yourself 2

Find the region satisfied by these inequalities.

$$x \geqslant 1 \qquad x + y < 5 \qquad y \geqslant 2x - 6$$

Solution

Test Yourself 2

$x = 1$ is the line through $(1, 0)$ and parallel to the y-axis.

$x + y = 5$ passes through $(0, 5)$ and $(5, 0)$ — A dotted line is used because = is not included.

$y = 2x - 6$ passes through $(0, -6)$ and $(2, -2)$ and $(4, 2)$.

$x \geqslant 1$	For $(0, 0)$, $x \geqslant 1$ is false. Shade the origin (unwanted) side.
$x + y > 5$	For $(0, 0)$, $x + y < 5$ is true. Shade the opposite (unwanted) side to the origin.
$y \geqslant 2x - 6$	For $(0, 0)$, $y \geqslant 2x - 6$ is true. Shade the opposite (unwanted) side to the origin.

The required region is labelled R.

8 Congruency

Congruent triangles

- For any two shapes to be congruent they must each contain the same angles and each have sides of the same lengths.

- To prove two triangles are congruent there are four different conditions, one of which must be satisfied:

  1 SSS: The three sides of one triangle are equal to the three sides of the other triangle.

   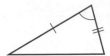

  2 SAS: Two sides and the included angle of one triangle are equal to two sides and the included angle of the other triangle.

  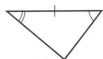 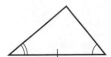

  3 ASA: Two angles and the side between them of one triangle are equal to two angles and the side between them of the other triangle.

  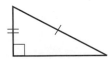 

  4 RHS: Each triangle is right-angled and the hypotenuse and one other side of one triangle are equal to the hypotenuse and one other side of the other triangle.

   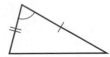

Chief Examiner says

- It is important that the order of the letters is correct when stating whether two triangles are congruent.
- Often when finding equal lengths or angles you will need to give a reason why they are equal.

Test Yourself 1

Are these triangles congruent?

If they are, use letters in the correct order to state clearly which triangles are equal. Give a reason for your answer. Also list the other sides and angles that are equal.

a

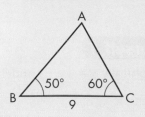 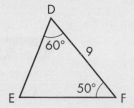

b

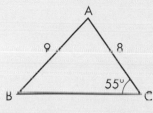 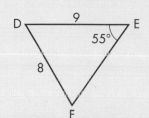

c

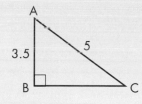 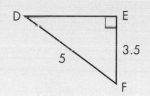

Test Yourself 2

Prove that the opposite sides of a parallelogram are equal.

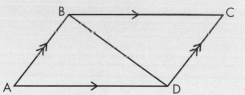

Solutions

Test Yourself 1

a Yes Triangle ABC is congruent to Triangle EFD (ASA)
Angle A = Angle E (third angle)
AB = EF (opposite 60°)
AC = ED (opposite 50°)

b No The sides that are equal are not in the corresponding position relative to the equal angle.

c Yes Triangle ABC is congruent to Triangle FED (RHS)
Angle A = Angle F
Angle C = Angle D
BC = ED

Test Yourself 2

In triangle ABD and triangle BDC:

Angle ABD = Angle BDC (alternate angles)

Angle ADB = Angle DBC (alternate angles)

> Angle ABD means the angle at B formed by lines from A and D.

Side DB is common to both triangles.

So triangle ABD is congruent to triangle CDB (ASA).

Therefore the other corresponding sides are equal.

So AB = CD and AD = CB (opposite sides are equal).

9 Simultaneous equations

Solving simultaneous equations graphically — Revised

- Make a table of values for each equation and plot the graphs on the same pair of axes.
- Write down the coordinates of the point(s) where the graphs cross.

Test Yourself 1

Solve graphically $y = 4x - 1$ and $x + y = 4$.

Solution

Test Yourself 1

$y = 4x - 1$

x	0	1	2
y	−1	3	7

$x + y = 4$

x	0	4	1
y	4	0	3

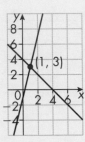

Answer: $x = 1$, $y = 3$

Solving simultaneous equations algebraically — Revised

- When solving simultaneous linear equations you are looking for a single value of x and a single value of y which satisfy both equations.

Method of elimination

- This is the best method if both equations are in the form $ax + by = c$.

    Step 1: Make the coefficients of x or y equal by multiplying one or both equations by a suitable number(s). Don't forget to multiply every term.

    Step 2: Eliminate the term with equal coefficients:

    If the signs of the 'equalised' terms are the same, subtract the equations.

    If the signs of the 'equalised' terms are the different, add the equations.

    This gives the solution for one of the variables.

    Step 3: Substitute this value into one of the equations to find the value of the other variable.

    Step 4: Write down both values as your answer.

- With harder equations you will have to multiply both equations by a different number to make the coefficients of one of the variables the same in both equations in Step 1.
- You can check your answer by substituting both values back into the equation not used in Step 3.

Test Yourself 2

Solve the following.

a $2x + 3y = 6$ (1)

$x + y = 4$ (2)

b $3x - 5y = 1$ (1)

$2x + 3y = 7$ (2)

Solutions

Test Yourself 2

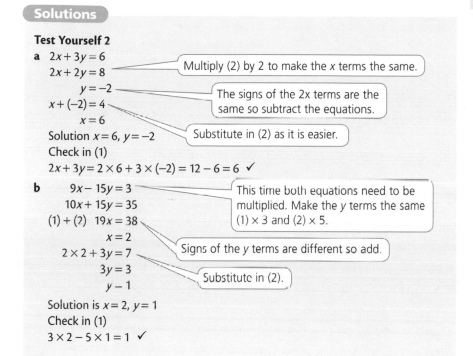

a $2x + 3y = 6$

$2x + 2y = 8$ ⟶ Multiply (2) by 2 to make the x terms the same.

$y = -2$ ⟶ The signs of the 2x terms are the same so subtract the equations.

$x + (-2) = 4$

$x = 6$

Solution $x = 6$, $y = -2$ ⟵ Substitute in (2) as it is easier.

Check in (1)

$2x + 3y = 2 \times 6 + 3 \times (-2) = 12 - 6 = 6$ ✓

b $9x - 15y = 3$ ⟶ This time both equations need to be multiplied. Make the y terms the same (1) × 3 and (2) × 5.

$10x + 15y = 35$

(1) + (2) $19x = 38$

$x = 2$

$2 \times 2 + 3y = 7$ ⟵ Signs of the y terms are different so add.

$3y = 3$

$y = 1$ ⟵ Substitute in (2).

Solution is $x = 2$, $y = 1$

Check in (1)

$3 \times 2 - 5 \times 1 = 1$ ✓

10 Vectors

The definition of a vector

- A vector is any quantity that has magnitude and direction.
- Vectors can be written as $\overrightarrow{AB}$ or **a**.

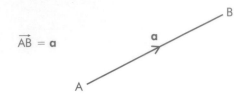

$\overrightarrow{AB} = \mathbf{a}$

- A vector on a grid can be written as the column vector $\begin{pmatrix} p \\ q \end{pmatrix}$, where p is the displacement in the x (horizontal) direction and q is the displacement in the y (vertical) direction. You use column vectors to describe translations.

Test Yourself 1

Write the column vectors representing

a $\overrightarrow{AB}$ **b** $\overrightarrow{CA}$ **c** $\overrightarrow{BC}$

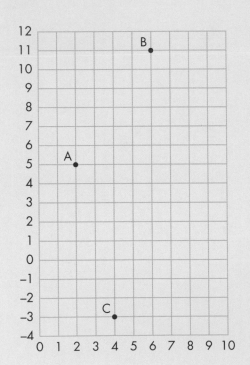

Solutions

Test Yourself 1

a $\begin{pmatrix} 4 \\ 6 \end{pmatrix}$ **b** $\begin{pmatrix} -2 \\ 8 \end{pmatrix}$ **c** $\begin{pmatrix} -2 \\ -14 \end{pmatrix}$

Multiplying a vector by a scalar
Revised

- $\mathbf{a} = k\mathbf{b}$ means that $\mathbf{a}$ is parallel to $\mathbf{b}$ and it is k times as long.
- $\overrightarrow{AB} = k\overrightarrow{AC}$ means that A, B and C are in a straight line and the length of AB is k times the length of AC.

Solutions

Test Yourself 2

a $3\begin{pmatrix} 3 \\ 5 \end{pmatrix} = \begin{pmatrix} 9 \\ 15 \end{pmatrix}$ ⟵ Multiply both numbers in the vector by 3.

b $\mathbf{c} = \begin{pmatrix} 6 \\ 10 \end{pmatrix} = 2\begin{pmatrix} 3 \\ 5 \end{pmatrix} = 2\mathbf{a}$

so **a** and **c** are parallel.

Test Yourself 2

$\mathbf{a} = \begin{pmatrix} 3 \\ 5 \end{pmatrix} \quad \mathbf{b} = \begin{pmatrix} 5 \\ 3 \end{pmatrix} \quad \mathbf{c} = \begin{pmatrix} 6 \\ 10 \end{pmatrix}$

a Work out 3**a**.

b Which of these vectors are parallel?

Addition and subtraction of vectors
Revised

- Vectors are added by starting the second vector where the first one finishes.
- To subtract two vectors, use $\mathbf{a} - \mathbf{b} = \mathbf{a} + -\mathbf{b}$.
- The resultant of two vectors **a** and **b** is $\mathbf{a} + \mathbf{b}$.
- To add column vectors, add the numbers on the top and add the numbers on the bottom separately: $\begin{pmatrix} a \\ b \end{pmatrix} + \begin{pmatrix} c \\ d \end{pmatrix} = \begin{pmatrix} a + c \\ b + d \end{pmatrix}$.
- Similarly, to subtract column vectors: $\begin{pmatrix} a \\ b \end{pmatrix} - \begin{pmatrix} c \\ d \end{pmatrix} = \begin{pmatrix} a - c \\ b - d \end{pmatrix}$.
- You can use vectors to prove geometrical results.

Test Yourself 3

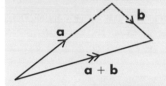

If $\mathbf{a} = \begin{pmatrix} 3 \\ 4 \end{pmatrix}$, $\mathbf{b} = \begin{pmatrix} 2 \\ -1 \end{pmatrix}$, find $\mathbf{a} + \mathbf{b}$.

Solution

Test Yourself 3

$\mathbf{a} + \mathbf{b} = \begin{pmatrix} 3 + 2 \\ 4 - 1 \end{pmatrix} = \begin{pmatrix} 5 \\ 3 \end{pmatrix}$

Solution

Test Yourself 4

$\overrightarrow{AB} = -\mathbf{a} + \mathbf{b}$

$\overrightarrow{AM} = \tfrac{1}{2}(-\mathbf{a} + \mathbf{b})$

$\overrightarrow{OM} = \overrightarrow{OA} + \overrightarrow{AM} = \mathbf{a} + \tfrac{1}{2}(-\mathbf{a} + \mathbf{b}) = \tfrac{1}{2}(\mathbf{a} + \mathbf{b})$

Test Yourself 4

In the diagram, $\overrightarrow{OA} = \mathbf{a}$ and $\overrightarrow{OB} = \mathbf{b}$. M is the midpoint of AB.

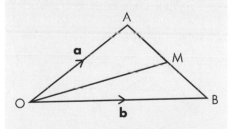

Find $\overrightarrow{OM}$ in terms of **a** and **b**.

11 Circle theorems

- All the points on a circle are the same distance from its centre.
- You need to know these terms.

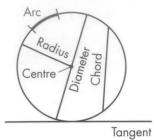

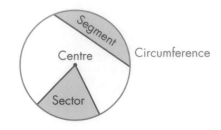

- The tangent at any point on a circle is at right angles to the radius at that point.
- The shaded parts are called the 'minor' sector and the 'minor' segment.
- The unshaded parts are called the 'major' sector and the 'major' segment.

- You will need to be able to apply the seven circle theorems and possibly show the proof of one of them.
- You need to be able to recognise which circle theorem(s) you need to use to solve a problem.
- The perpendicular from the centre of a circle to a chord bisects the chord.
- The angle subtended by an arc at the centre is twice the angle subtended at the circumference.

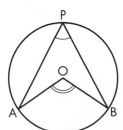

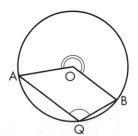

Angle AOB = 2 × Angle APB   Reflex angle AOB = 2 × Angle AQB

- The angle in a semicircle is a right angle.

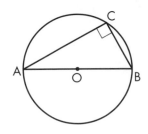

Angle ACB = 90°

- Angles in the same segment are equal.

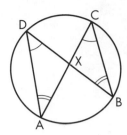

Angle ADB = Angle ACB
and Angle DAC = Angle DBC

- Opposite angles of a cyclic quadrilateral add up to 180°.

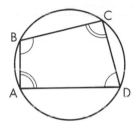

Angle ABC + Angle CDA = 180°
and Angle BAD + Angle BCD = 180°

- The two tangents to a circle from a point outside a circle are equal in length.

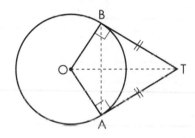

- Angles in the alternate segment are equal.

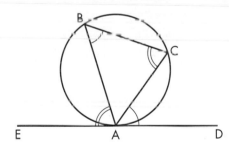

Angle ABC = Angle CAD
and Angle BAE = Angle ACB

Test Yourself 1

a In Figure 1, TA = 9 cm and TO = 10 cm.
Find the radius of the circle.

b In Figure 1, Angle ATB = 50°.
Find Angle TAB.

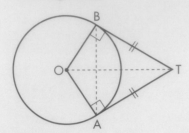

Figure 1

e In Figure 4, Angle CAO = 38˚ and
O is the centre of the circle.
Find Angle CBA.

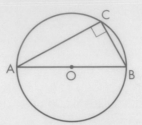

Figure 4

c In Figure 2, Angle AOB = 120° and
the radius of the circle = 5 cm.
Find AB.

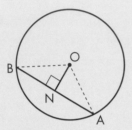

Figure 2

f In Figure 5, Angle BCD = 102˚.
Find Angle BAD.
Give your reasons.

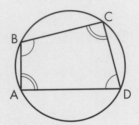

Figure 5

d In Figure 3, Angle AOB = 68˚.
Find Angle APB.

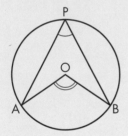

Figure 3

Solutions

Test Yourself 1

a Since Angle TAO = 90°, use Pythagoras.
$OA^2 + 9^2 = 10^2$
$OA = \sqrt{100 - 81} = 4.36$ cm

b Since TA = TB, triangle TAB is isosceles.
Angle TAB = $\frac{1}{2}(180 - 50) = 65°$

c Angle BON = $\frac{1}{2}(120°) = 60°$
$\frac{BN}{OB} = \sin 60°$, so BN = $5 \times \sin 60° = 4.33$ cm
Since AN = BN, AB = $4.33 \times 2 = 8.66$ cm

d Angle APB = $\frac{1}{2}$ Angle AOB = $\frac{1}{2} \times 68 = 34°$

e Angle ACB = 90°
So Angle CBA = $180 - 38 - 90 = 52°$

f Angle BAD = $180 - 102 = 78°$ because Angles BCD and BAD are opposite angles of a cyclic quadrilateral.

Chief Examiner says

When giving reasons, use the standard phrases, e.g.
'angles in a semicircle',
'angle at centre = twice angle at circumference',
'angles in the same segment',
'opposite angles of a cyclic quadrilateral'.

12 Scatter diagrams and time series

Scatter diagrams

- You use scatter diagrams (or graphs) to find out if there is a link between two variables.
- The values of the two variables are plotted as points on a graph.
- If the points lie approximately in a straight line, then there is a correlation between the two variables.
- The closer the points are to a straight line, the stronger the correlation.
- The correlation is positive if the higher the value of x, the higher the value of y.
- The correlation is negative if the higher the value of x, the lower the value of y.
- If there is a correlation, then you can draw a line of best fit. It should reflect the slope of the points and have approximately the same number of points on either side.
- You can use a line of best fit to estimate a value of y from a given value of x or vice versa.
- You should not use a line of best fit to estimate too far outside the range of the data.

Positive correlation

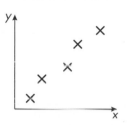

Negative correlation

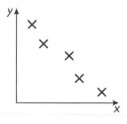

Test Yourself 1

The table shows the time spent and marks gained by 10 students for their coursework.

a Plot a scatter diagram for this information.

b What conclusions can be drawn?

c Draw a line of best fit.

d Use your line of best fit to estimate the mark of a student who spent nine hours on the coursework.

Time (h)	5	8	3	6	6	7	4	10	8	7
Mark	12	15	9	14	11	13	10	19	16	14

Solutions

Test Yourself 1

a c

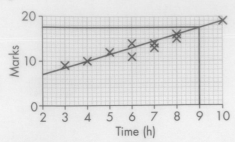

b Fairly strong positive correlation

d 17 or 18

Time series

Revised

- A time series is a graph with time on the horizontal axis and any quantity which varies with time on the vertical axis. For example, quarterly sales of a company.

- You can use a moving average when the graph shows a cyclical change. For example, higher sales always in the summer and lower sales always in the winter.

- To calculate the moving average in the example of quarterly sales:
  - Find the mean for the four quarters in the first year.
  - Omit the first quarter of year one and include the first quarter of year two and find the new mean.
  - Omit the second quarter of year one and include the second quarter of year two and so on.

- The moving average is plotted in the middle of the four quarters.

Test Yourself 2

The table gives the sales figures in £millions for a company.

Quarter	1st	2nd	3rd	4th
2008	4.2	5.6	7.2	4.9
2009	3.9	5.5	6.8	4.7
2010	3.8	5.2	6.5	4.6

a Plot a time series graph.

b Calculate the four-quarter moving averages and plot them on the graph.

c Comment on the seasonal variation and the long-term trend.

Test Yourself 2

a

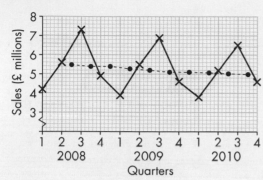

b (4.2 + 5.6 + 7.2 + 4.9) ÷ 4 = 5.475
(5.6 + 7.2 + 4.9 + 3.9) ÷ 4 = 5.4
(7.2 + 4.9 + 3.9 + 5.5) ÷ 4 = 5.375
(4.9 + 3.9 + 5.5 + 6.8) ÷ 4 = 5.275
(3.9 + 5.5 + 6.8 + 4.7) ÷ 4 = 5.225
(5.5 + 6.8 + 4.7 + 3.8) ÷ 4 = 5.2
(6.8 + 4.7 + 3.8 + 5.2) ÷ 4 = 5.125
(4.7 + 3.8 + 5.2 + 6.5) ÷ 4 = 5.05
(3.8 + 5.2 + 6.5 + 4.6) ÷ 4 = 5.025

c Sales are highest in the third quarter and lowest in the first.
The long-term trend is for slightly falling sales.

1 Algebraic manipulation

Expanding brackets in algebra

- When expressions of the form $(a + b)(c + d)$ are expanded, every term in the first bracket is multiplied by every term in the second bracket.
  $(a + b)(c + d) = ac + ad + bc + bd$
- You may have various strategies for organising this work, for example 'FOIL' (firsts, outsides, insides, lasts) or using a multiplication table.
- Having expanded the brackets to four terms, you can often combine two 'like terms' to simplify the expression.

> **Chief Examiner says**
>
> 'Multiply out', 'Expand' and 'Remove the brackets' all mean the same thing.

Test Yourself 1

Multiply out the brackets in the following.

a $(x + 4)(x + 2)$

b $(3a + 5)(4a - 3)$

c $(x - 3)^2$

Solutions

Test Yourself 1

a $x^2 + 2x + 4x + 8 = x^2 + 6x + 8$

b $12a^2 - 9a + 20a - 15 = 12a^2 + 11a - 15$

c $(x - 3)(x - 3) = x^2 - 3x - 3x + 9 = x^2 - 6x + 9$

> Multiply the bracket by itself.

Surds

- You can use algebraic methods to manipulate expressions involving surds.

Test Yourself 2

Simplify these.

a $(2 + \sqrt{3})(5 + 2\sqrt{3})$

b $(5 + \sqrt{2})^2$

Solutions

Test Yourself 2

a $(2 + \sqrt{3})(5 + 2\sqrt{3})$
   $= 10 + 4\sqrt{3} + 5\sqrt{3} + 2\sqrt{3}\sqrt{3}$
   $= 10 + 9\sqrt{3} + 6 = 16 + 9\sqrt{3}$

> $2 \times \sqrt{3} \times \sqrt{3} = 2 \times 3 = 6$

b $(5 + \sqrt{2})^2 = 25 + 5\sqrt{2} + 5\sqrt{2} + \sqrt{2} \times \sqrt{2}$
   $= 25 + 10\sqrt{2} + 2$
   $= 27 + 10\sqrt{2}$

2 Perimeter, area, volume and 2-D representation

The area of a rectangle and of a triangle
Revised

- Area of a rectangle
  $A = b \times h$

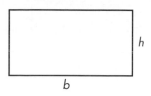

- Area of a triangle
  $A = \dfrac{h \times h}{2}$

Test Yourself 1

The vertices of a triangle are at (2, 1), (2, −3) and (14, −3).

Draw axes and plot the triangle. Find its area.

Solution

Test Yourself 1

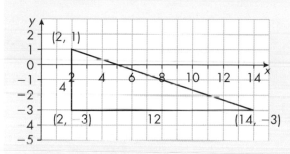

$A = \dfrac{12 \times 4}{2} = 24 \text{ units}^2$

The area of a parallelogram
Revised

- Area of a parallelogram
  $A = b \times h$

Test Yourself 2

Find the area of this parallelogram.

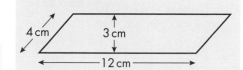

Solution

Test Yourself 2
$A = b \times h$

$= 12 \times 3$ ──── Make sure you use the perpendicular height, not the sloping side.

$= 36 \text{ cm}^2$

The area of complex shapes

- You need to learn the formulae for
  - the area of a rectangle
  - the area of a triangle
  - the area of a parallelogram.
- You may have to combine them to find the area of a complex shape.

Test Yourself 3

The diagram shows the end of a shed.

Find the area of the end of the shed.

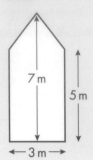

Solution

Test Yourself 3

Area of rectangle $= 5 \times 3$

> The shape is made up of a rectangle of height of 5 m and base of 3 m ...

$= 15\,m^2$

Area of triangle $= \dfrac{3 \times 2}{2} = 3\,m^2$

> ... and a triangle with a base of 3 m and a height of 7 m − 5 m = 2 m.

Area of end of shed $= 15 + 3 = 18\,m^2$

The circumference of a circle and The area of a circle

- Area of a circle
  $A = \pi r^2$

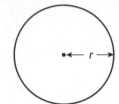

- Circumference of a circle
  $C = \pi d$

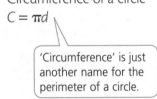

> 'Circumference' is just another name for the perimeter of a circle.

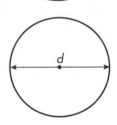

- You need to learn these formulae.
- You may be asked to find an exact value for the circumference or area of a circle. This means leave π in your answer.
- You may have to combine them with the formulae for straight-sided shapes to deal with complex shapes.

Test Yourself 4

A semicircular flowerbed has a diameter of 8 m.

Find the perimeter of the flowerbed in terms of π.

Solutions

Test Yourself 4

Circumference of semicircle $= \dfrac{\pi d}{2}$

> Do not substitute the numerical value of π.

$= \dfrac{\pi \times 8}{2} = 4\pi$

> Do not forget the straight edge of the semicircle.

Perimeter of the flowerbed $= 4\pi + 8$ m

Test Yourself 5

$A = \pi r^2$

$40 = \pi \times r^2$

$\dfrac{40}{\pi} = r^2$

$r = \sqrt{\dfrac{40}{\pi}}$

$r = 3.6$ cm (to 1 d.p.)

Test Yourself 6

a Diameter of circle = 10 cm, so radius = 5 cm

Area of semicircle $= \frac{1}{2} \times \pi \times 5^2$

$\qquad\qquad\qquad = 39.269 \dots$ cm^2

Area of parallelogram $= 10 \times 4$

$\qquad\qquad\qquad\qquad = 40$ cm^2

Area of whole shape = 79.3 cm^2 (1 d.p.)

b Perimeter of shape

$= 5 + 10 + 5 + \frac{1}{2} \times \pi \times 10$

$= 35.7$ cm (1 d.p.)

Test Yourself 5

A circular pond has an area of 40 m^2.

Calculate the radius of the pond.

Test Yourself 6

This shape is made from a parallelogram and a semicircle.

a Find its area.

b Find its perimeter.

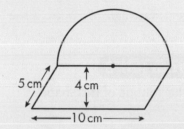

5 cm   4 cm

10 cm

The volume of complex shapes

Revised

- Volume of a cuboid

  $V = l \times w \times h$

- You can find the volume of complex shapes by splitting them into shapes you can find the volume of.

Test Yourself 7

Volume = Volume of large cuboid − Volume of small cuboid

$= 10 \times 5 \times 4 - 3 \times 2 \times 2$

$= 200 \quad 12$

$= 188$ cm^3

Test Yourself 7

Find the volume of this shape.

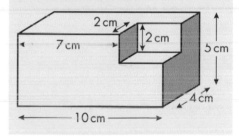

2 cm

7 cm   2 cm

5 cm

10 cm   4 cm

Chief Examiner says

Don't forget to include the units in your answer.

The volume of a prism

- A prism is a three-dimensional shape with a constant cross-section.
- These are some examples of prisms.

  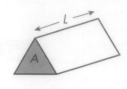

Cuboid   Cylinder   Triangular prism

- For all prisms, Volume = Area of cross-section × length, or $V = A \times L$.
- The surface area of a prism is the total area of all the surfaces of the prism.

Test Yourself 8

Find the volume of this shape.

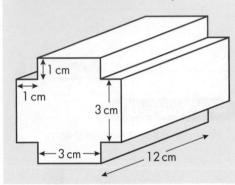

Solution

Test Yourself 8

Area of cross-section $= 5 \times 5 - 4 \times 1 \times 1$

$\qquad\qquad\qquad\qquad\quad = 25 - 4$

$\qquad\qquad\qquad\qquad\quad = 21\,\text{cm}^2$

Volume of prism = Area of cross-section × length

$\qquad\qquad\qquad\quad = 21 \times 12$

$\qquad\qquad\qquad\quad = 252\,\text{cm}^3$

The volume of a cylinder and The surface area of a cylinder

- Volume of a cylinder
  $V = \pi r^2 h$

- Curved surface area of a cylinder $= 2\pi r h$

- Total surface area of a cylinder $= 2\pi r h + 2\pi r^2$

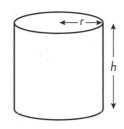

Chief Examiner says

Don't use rounded answers in the middle of a calculation. Keep all the figures on your calculator (e.g. use the answer function) and just round at the end. Give your answer to a sensible degree of accuracy if no specific accuracy is stated.

Test Yourself 9

Find the volume and surface area of this packet of sweets.

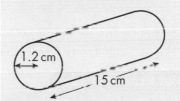

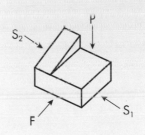

Solution

Test Yourself 9

Volume $= \pi r^2 h$

$\quad = \pi \times 1.2^2 \times 15$

$\quad = 67.9\,cm^3$ (1 d.p.)

Total surface area $= 2\pi r h + 2\pi r^2$

$\quad = 2 \times \pi \times 1.2 \times 15 + 2 \times \pi \times 1.2^2$

$\quad = 122.1\,cm^2$ (1 d.p.)

Plans and elevations

- The plan view of an object is the view seen directly from above.

- The front elevation is the view seen directly from the front.

- The side elevation is the view seen directly from one side (there are two possible side elevations).

Chief Examiner says

Remember that hidden edges are shown as dotted lines and right angles should be marked as right angles.

Test Yourself 10

Sketch the plan (P), the front elevation (F) and side elevations (S_1 and S_2) of this shape.

Solution

Test Yourself 10

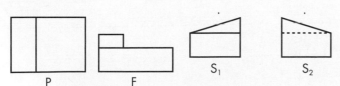

3 Trial and improvement

Solving equations by trial and improvement

- Some equations cannot be solved by algebraic methods. To solve them you need to use a trial and improvement procedure.

- To start trial and improvement you need a first estimate. This could come from a graph, but in examinations you are usually given a hint to the first estimate.

- Start with two values, one which gives too big a result and the other too small. The true solution will be between them.

- Try in between (usually half way).

- Continue the process, finding two values one of which is too big and the other too small until you have an answer to the required accuracy.

Test Yourself 1

a Show that there is a solution to $x^3 + 4x = 8$ which lies between 1 and 2.

b Find the solution correct to 1 decimal place.

Chief Examiner says

Don't forget to write down the answer to each trial.

It is often best to work in a table.

Solutions

Test Yourself 1

a

x	$x^3 + 4x$	Too big / Too small
1	5	Too small
2	16	Too big
b 1.5	9.375	Too big
1.2	6.528	Too small
1.3	7.397	Too small
1.4	8.344	Too big
1.35	7.860375	Too small

These show that x lies between 1 and 2.

So x is between 1 and 1.5.

So x is between 1.2 and 1.5.

So x is between 1.3 and 1.4.
So correct answer to 1 d.p. is either 1.3 or 1.4. To check which is correct try half way between.

The answer is between 1.35 and 1.4, so $x = 1.4$ correct to 1 decimal place.

Chief Examiner says

Don't forget in a question like this that you are trying to find x to 1 decimal place (1.4). You are not trying to get the value of $x^3 + 4x$ to be 8 to 1 decimal place.

4 Probability 1

Revised

Basic probability

- Probabilities can be expressed as fractions, decimals or percentages. All probabilities are between 0 and 1 inclusive.
- P(A does not happen) = $1 - P(A)$.

> P(A) is a useful shorthand for the probability that A happens.

- For equally likely outcomes

$$P(A) = \frac{\text{Number of ways A can happen}}{\text{Total number of possible outcomes}}$$

- Mutually exclusive outcomes are those which cannot happen together. If A, B and C are mutually exclusive outcomes covering all the possibilities, then $P(A) + P(B) + P(C) = 1$.

Test Yourself 1

A jar contains red, green and white beads. One bead is selected at random.

The probability that the bead is red is 0.3.

The probability that the bead is green is 0.5.

What is the probability that the bead is white?

Solution

Test Yourself 1

$P(R) + P(G) = 0.3 + 0.5 = 0.8$

$P(W) = 1 - 0.8 = 0.2$

Chief Examiner says

Never write probabilities as '1 in 5' or as a ratio '1 to 5' or 1 : 5. You will lose marks if you do. Instead, write $\frac{1}{5}$ or 0.2 (or 20% if the question uses percentages).

Expected frequency

Revised

- Expected frequency = Probability × Number of trials

Solutions

Test Yourself 2

a $\frac{1}{6} \times 120 = 20$

b $\frac{3}{6} \times 120 = 60$

Test Yourself 2

A fair dice is rolled 120 times.

How many times would you expect it to show

a a six

b an odd number?

Relative frequency

Revised

- Relative frequency = $\dfrac{\text{Number of times event occurs}}{\text{Total number of trials}}$

- When theoretical probabilities are not known, relative frequency can be used to estimate probability.

- The greater the number of trials, the better the estimate. The graph of relative frequency against the number of trials may vary greatly at first, but later 'settles down'.

- Relative frequency experiments may be used to test for bias, for example to see if a dice is fair.

Test Yourself 3

Sarah has a biased dice. This graph shows the relative frequency of throwing a 6 when she threw the dice 1000 times.

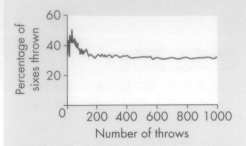

Use the graph to estimate the probability of throwing a 6 with Sarah's dice. Explain your answer.

Solution

Test Yourself 3

Since Sarah has thrown the dice a large number of times she can use the result from 1000 throws as an estimate of the probability of throwing a 6.

From the graph this is about 30% or 0.3.

5 Graphs 1

Real-life graphs
Revised

- Look at the labels on the axes – they tell you what the graph is about.
- Look whether the graph is a straight line or a curve.
- The slope of the graph gives you the rate of change.
- For distance–time graphs, the rate of change is the velocity.

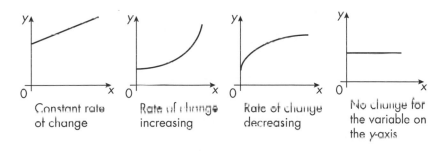

| Constant rate of change | Rate of change increasing | Rate of change decreasing | No change for the variable on the y-axis |

Test Yourself 1

Water is poured into this vessel at a constant rate.

Sketch a graph of depth of water (d cm) against time (t secs).

Solution

Test Yourself 1

Since the radius is decreasing, at first, the rate of change will increase. At the top the radius is constant and so the rate of change will be constant.

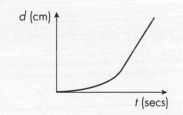

Distance–time graphs and Velocity–time graphs
Revised

- The gradient (rate of change) on a distance–time graph gives the velocity.
- If the rate of change is constant, the velocity is constant.
- In a velocity–time graph the gradient gives the acceleration, or deceleration if negative.

Solution

Test Yourself 2

During first second, the object accelerates from 0 to 9 m/s.

It continues at a constant velocity of 9 m/s for 2 seconds.

It then decelerates over next 2 seconds to 0 m/s.

Test Yourself 2

Describe what is happening in this velocity–time graph.

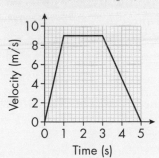

Quadratic graphs

- The equations of quadratic graphs are in the form
  $y = ax^2 + bx + c$.

- The shape of quadratic graphs is

for $a > 0$ for $a < 0$

- The name of this shape graph is a parabola. All parabolas are symmetrical. This symmetry can be seen on the graph and can often be seen in the table as well.

- These are the steps to draw a quadratic graph.
  - Make a table of values. You may want extra rows in your table between the x and y rows.
  - Plot the points. Remember the curve will be symmetrical.
  - Join the points with a smooth curve.

> **Chief Examiner says**
>
> Don't forget to join the points. You will lose marks if you do not do so. Also it may be impossible to answer any questions about the graph if you do not draw the curve.

- You can use graphs of quadratic functions to solve quadratic equations.
  - Plot the graph of $y = ax^2 + bx + c$.
  - To solve the equation $ax^2 + bx + c = 0$, read off the values of x where the curve crosses the x axis ($y = 0$).
  - To solve the equation $ax^2 + bx + c = k$, draw the line $y = k$ and read off the values of x where the curve crosses the line.

> **Chief Examiner says**
>
> Don't forget the solutions to $ax^2 + bx + c = k$ are x values not y values. You may lose marks if you include the y coordinates of the points where the curve crosses the line.

- Most quadratic equations have two possible solutions.

> **Test Yourself 3**
>
> **a** Draw the graph of
> $y = x^2 - 4x + 3$
> for $-1 \leqslant x \leqslant 5$.
>
> **b** Use your graph to solve these equations.
>   **i** $x^2 - 4x + 3 = 0$
>   **ii** $x^2 - 4x + 3 = 6$

Test Yourself 3

a

x	−1	0	1	2	3	4	5
x^2	1	0	1	4	9	16	25
$-4x$	4	0	−4	−8	−12	−16	−20
$+3$	3	3	3	3	3	3	3
$y - x^2 - 4x + 3$	8	3	0	−1	0	3	8

> For example when $x = -1$
> $y = (-1)^2 - 4(-1) + 3$
> $\quad = 1 + 4 + 3 = 8$

and

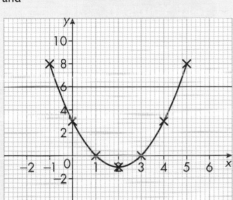

b i Look where the graph crosses the x-axis ($y = 0$): the solutions are $x = 1$ and $x = 3$.

ii Draw the line $y = 6$ on your graph.

Look where the graph crosses the line $y = 6$: the solutions are $x = -0.6$ and $x = 4.6$ correct to 1 d.p.

6 Measures

Converting between measures Revised

- As well as changing between linear metric units, you need to know how to change area and volume units.

  Length: $1\,cm = 10\,mm$ $\qquad\qquad$ $1\,m = 100\,cm$

  Area: $\quad 1\,cm^2 = 10^2\,mm^2 = 100\,mm^2$ $\quad 1\,m^2 = 100^2\,cm^2 = 10\,000\,cm^2$

  Volume: $1\,cm^3 = 10^3\,mm^3 = 1000\,mm^3$ $\;\, 1\,m^3 = 100^3\,cm^3 = 1\,000\,000\,cm^3$

Chief Examiner says

If you are asked to change the units of area or volume, it is often easiest to change the lengths you are given to the new units and to work out the area or volume using those. If you are not given the lengths, you need to use the information in the list above.

Test Yourself 1

Change these units.

a $10\,500\,cm^2$ to m^2

b $4.2\,cm^3$ to mm^3

Solutions

Test Yourself 1

a $\qquad 10\,000\,cm^2 = 1\,m^2$

$\quad$ so $10\,500\,cm^2 = 10\,500 \div 10\,000$

$\qquad\qquad\qquad\quad = 1.05\,m^2$

b $\qquad 1\,cm^3 = 1000\,mm^3$

$\quad$ so $4.2\,cm^3 = 4.2 \times 1000$

$\qquad\qquad\quad = 4200\,mm^3$

Accuracy of measurement Revised

- A length, x, measured as $45\,cm$ to the nearest centimetre lies between $44.5\,cm$ and $45.5\,cm$.

- You can write this as $44.5 \leqslant x < 45.5\,cm$.

- The lower bound is $44.5\,cm$ and the upper bound is $45.5\,cm$.

Chief Examiner says

A common error is to give the upper bound as $45.49\,cm$ in an example like this, since the upper bound cannot be reached. The correct bound, however, is $45.5\,cm$ here.

Test Yourself 2

Give the bounds of these measurements.

a $86\,mm$ to the nearest millimetre

b $48\,g$ to the nearest gram

c 52 litres to the nearest litre

d $80\,cm$ to the nearest centimetre

Solutions

Test Yourself 2

a $85.5\,mm$ and $86.5\,mm$

b $47.5\,g$ and $48.5\,g$

c 51.5 litres and 52.5 litres

d $79.5\,cm$ and $80.5\,cm$

Working to a sensible degree of accuracy Revised

- Sometimes you need to work to a reasonable degree of accuracy. To decide what is reasonable, look at the context and the accuracy of the information you have been given. Your answer should not be more accurate than the figures you used to calculate it.

Compound measures

- The following are examples of compound measures.

  - Speed $= \dfrac{\text{Distance}}{\text{Time}}$

  - Density $= \dfrac{\text{Mass}}{\text{Volume}}$

  - Population density $= \dfrac{\text{Population}}{\text{Area}}$

Chief Examiner says

The units tell you which way to divide. Density measured in g/cm³ means grams (mass) divided by cm³ (volume).

Test Yourself 3

Jane drives for 2 hours at 70 mph and then for 30 minutes at 40 mph.

a Find the total distance for the journey.

b Calculate her average speed for the journey.

Solutions

Test Yourself 3

a Distance = Speed × Time

At 70 mph she travels $70 \times 2 = 140$ miles.

At 40 mph she travels $40 \times 0.5 = 20$ miles.

Total distance $= 140 + 20$

$\qquad\qquad\qquad = 160$ miles

30 minutes = 0.5 hours

b Total time = 2.5 hours

Average speed $= \dfrac{\text{Total distance}}{\text{Total time}}$

$\qquad\qquad\quad = \dfrac{160}{2.5}$

$\qquad\qquad\quad = 64$ mph

Chief Examiner says

Don't be tempted to find the average of the two speeds – the times are different.

More on bounds of measurement and Further calculations involving bounds of measurements

- Measurements are not always given correct to the nearest whole unit. For measurements given correct to the nearest part of a unit, the possible error is half of the part of the unit. For example, the upper and lower bounds of 4.5 cm correct to 1 decimal place (i.e. to the nearest 0.1 cm) are 4.55 and 4.45 cm.

- To find the upper bound of a sum, add the two upper bounds.

- To find the lower bound of a sum, add the two lower bounds.

- To find the upper bound of a difference, subtract the lower bound from the upper bound.

- To find the lower bound of a difference, subtract the upper bound from the lower bound.

- To find the upper bound of a product, multiply the upper bounds.

- To find the lower bound of a product, multiply the lower bounds.

- To find the upper bound of a division, divide the upper bound by the lower bound.
- To find the lower bound of a division, divide the lower bound by the upper bound.

Solutions

Test Yourself 4

a Upper bound = $10 \times 5.5 = 55$ kg
   Lower bound = $10 \times 4.5 = 45$ kg

b i Minimum perimeter = $2 \times 3.45 + 2 \times 4.55 = 16$ cm
   Maximum perimeter = $2 \times 3.55 + 2 \times 4.65 = 16.4$ cm

   ii Minimum area = $3.45 \times 4.55 = 15.6975 = 15.7$ cm^2
   Maximum area = $3.55 \times 4.65 = 16.5075 = 16.5$ cm^2

c Lower bound of $A = \dfrac{3 \times 3.615}{5.415} = 2.0028 = 2.00$

   Upper bound of $A = \dfrac{3 \times 3.625}{5.405} = 2.0120 = 2.01$

a A box of apples has a mass of 5 kg to the nearest kilogram. What are the upper and lower bounds of the mass of 10 of these boxes?

b A rectangle has sides 3.5 cm and 4.6 cm measured to 2 significant figures. Find the minimum and maximum value of the following.

   i The perimeter

   ii The area

c $A = \dfrac{3b}{c}$, $b = 3.62$, $c = 5.41$

   to 2 d.p.

Find the lower and upper bounds of A.

7 Percentage and proportional change

Revised

Solving problems

- When solving a problem, break it down into steps.
- Ask yourself these questions.
  - What am I asked to find?
  - What information have I been given?
  - What methods can I apply?

Index numbers

- Index numbers are used to show how much something has increased (or decreased) over time.
- When the system is started the index is taken as 100. This means it is 100% of the value at that time. If the value increases by, say, 15%, the new index will be 115.

Test Yourself 1

In March 2009, the CPI (Consumer Prices Index) was 109.8. In March 2011 it had risen to 118.1.

Calculate the percentage increase in these prices.

Solution

Test Yourself 1

Increase in index = 118.1 − 109.8

$\qquad = 8.3$

Percentage increase $= \frac{8.3}{109.8} \times 100$

$\qquad = 7.559... = 7.6\%$

Repeated percentage change

Revised

- To calculate a percentage increase or decrease using a multiplier, see Unit B Chapter 2.
- To calculate repeated percentage change, use the multiplier raised to a power. The power is the number of times you want to repeat the change. For example, to increase by 6% five times, multiply by $(1.06)^5$.

Chief Examiner says

- To check how to work out $(1.06)^5$ on your calculator, see Unit A Chapter 1.
- Compound interest is repeated percentage increase.

Test Yourself 2

a Sheila invests £6000 at 4% compound interest. How much will the investment be worth after 10 years?

b A car depreciates in value by 12% per year. It cost £12500 when new. How much will it be worth after

   i 2 years   **ii** 5 years?

Solutions

Test Yourself 2

a Value $= 6000 \times (1.04)^{10}$

$\qquad = £8881.47$ (to the nearest penny)

b i Value after 2 years $= 12\,500 \times 0.88^2$

$\qquad\qquad = £9680$

ii Value after 5 years $= 12\,500 \times 0.88^5$

$\qquad\qquad = £6596.65$ (to the nearest penny)

Repeated fractional change
Revised

- To increase or decrease an amount by a fraction:
  - Add or subtract the fraction to or from 1.
  - Multiply by the new fraction.
- To calculate repeated fractional change, use the multiplier raised to a power. The power is the number of times you want to repeat the change. For example, to increase by $\frac{1}{4}$ five times, multiply by $(\frac{5}{4})^5$.

Test Yourself 3

Increase 18 by $\frac{1}{3}$.

Test Yourself 4

An antique painting is bought for £2000.

Its value increases by $\frac{1}{6}$ each year.

How much is the painting worth after 5 years?

Solutions

Test Yourself 3

$18 \times \frac{4}{3} = 24$ $\quad$ $1 + \frac{1}{3} = \frac{4}{3}$

Test Yourself 4

$2000 \times (\frac{7}{6})^5 = £4323$ (to the nearest pound)

Reverse percentage and fraction problems
Revised

- To find the original amount before a percentage change,
  - Add or subtract the percentage to or from 100
  - Change to a decimal
  - Divide.
- To find the original amount before a fractional change,
  - Add or subtract the fraction to or from 1.
  - Divide by the fraction.

Test Yourself 5

The price of a skirt was reduced by 5%. It now costs £27.55. What was the original price?

Test Yourself 6

After an increase of $\frac{1}{5}$, the cost of a coat is £54.

What did it cost before the increase?

Solutions

Test Yourself 5

Decrease, so new price $= 100 - 5$

$\qquad = 95\%$ of original price.

Original price $= 27.55 \div 0.95 = £29.$ $\quad$ To find original price, divide by 0.95 (95% = 0.95).

Test Yourself 6

Increase, so new price $= 1 + \frac{1}{5}$

$\qquad = \frac{6}{5}$ of the original price.

Original price $= £54 \div 1\frac{1}{5}$

$\qquad = £54 \div \frac{6}{5}$

$\qquad = £54 \times \frac{5}{6} = £45$

8 Standard form and using a calculator

Standard form and Calculating with numbers in standard form — Revised

- This is used for very large and very small numbers.
- They are written in the form $a \times 10^n$, where n is an integer and $1 \leqslant a < 10$
- To multiply or divide numbers in standard form without a calculator, use the rules of indices. For example
  $(3 \times 10^6) \times (5 \times 10^8) = 15 \times 10^{6+8} = 15 \times 10^{14} = 1.5 \times 10^{15}$
- To add or subtract numbers in standard form without a calculator, it is safer to change the numbers to ordinary numbers first.

Standard form on your calculator

- Use the $\boxed{\text{EXP}}$ or $\boxed{\text{EE}}$ or $\boxed{\times 10^x}$ key.
- Use this key in place of the $\times 10$ part of a standard form number. For example, to enter 3.5×10^6 press

$\boxed{3}\boxed{.}\boxed{5}\boxed{\text{EXP}}\boxed{6}\boxed{=}$

Solutions

Test Yourself 1

a i 3.5×10^6
  ii 4.2×10^{-4}
  iii 4.32×10^1
  iv 3×10^{-2}

b i 4000
  ii 27 400
  iii 0.53
  iv 0.000 468

c i $28 \times 10^{8+3} = 28 \times 10^{11} = 2.8 \times 10^{12}$
  ii $3 \times 10^{7-4} = 3 \times 10^3$
  iii $42 \times 10^{6+-2} = 42 \times 10^4 = 4.2 \times 10^5$
  iv $83\,000 + 9800 = 92\,800 = = 9.28 \times 10^4$

> Give your answer in standard form

d i  $\boxed{2}\boxed{.}\boxed{3}\boxed{\text{EXP}}\boxed{6}\boxed{+}\boxed{5}\boxed{\text{EXP}}\boxed{7}\boxed{=}$ 5.23×10^7

  ii $\boxed{[}\boxed{6}\boxed{.}\boxed{2}\boxed{\text{EXP}}\boxed{8}\boxed{-}\boxed{7}\boxed{.}\boxed{5}\boxed{\text{EXP}}\boxed{7}\boxed{]}$

     $\boxed{\div}\boxed{1}\boxed{.}\boxed{5}\boxed{\text{EXP}}\boxed{3}\boxed{=}$ 3.64×10^5 (to 3 s.f.)

Chief Examiner says

Use proper standard form notation in your answer. Don't just write down your calculator display.

Test Yourself 1

a Write these numbers in standard form.
  i 3 500 000
  ii 0.000 42
  iii 43.2
  iv 0.03

b Write these as ordinary numbers.
  i 4×10^3
  ii 2.74×10^4
  iii 5.3×10^{-1}
  iv 4.68×10^{-4}

c Work out these without a calculator.
  i $4 \times 10^8 \times 7 \times 10^3$
  ii $9 \times 10^7 \div 3 \times 10^4$
  iii $7 \times 10^6 \times 6 \times 10^{-2}$
  iv $8.3 \times 10^4 + 9.8 \times 10^3$

d Use your calculator to work out the following.
  i $2.3 \times 10^6 + 5 \times 10^7$
  ii $\dfrac{6.2 \times 10^8 - 7.5 \times 10^7}{1.5 \times 10^3}$

The efficient use of a calculator

> **Chief Examiner says**
>
> Calculators vary. The position and symbols used on the buttons differ. Make sure you know how **your** calculator works. Don't borrow a different calculator or change your calculator just before an exam.

- You need to be able to use your calculator efficiently.
- Make sure you can use your calculator to work with these.
  - Negative numbers
  - Reciprocals
  - Powers and roots
  - The trigonometry functions: sine, cosine and tangent
  - Standard form
- You also need to be able to use brackets to ensure your calculator performs calculations in the correct order.

Test Yourself 2

Work out $\dfrac{4 + \tan 45°}{5^2}$

Solution

Test Yourself 2

 $= 0.2$

Use brackets to ensure the addition is done before the division.

Exponential growth and decay

- Use a constant multiplier – greater than 1 for growth, less than 1 for decay.
- The formula for exponential growth or decay is $y = A \times b^x$ where A is the initial value and b is the rate of growth or decay.
- Compound interest and depreciation are examples of exponential growth and exponential decay.
- Use the $\boxed{\wedge}$ or $\boxed{x^y}$ key.

Test Yourself 3

a A car depreciates by 30% every year. It cost £9000 new. How much is it worth after 5 years?

b A population of bacteria increases by 6% every hour. By what factor has the population grown after 24 hours?

Solutions

Test Yourself 3

a $9000 \times (0.7)^5$
 $= £1512.63$

b $(1.06)^{24} = 4.0489...$
 Approximately 4 times.

9 Similarity

Similar shapes

Revised

- For two shapes to be similar, one shape is an enlargement of the other so
  - all corresponding sides must have proportional lengths and
  - all corresponding angles must be equal.
- To prove that two triangles are similar you only need to prove that one of these conditions is true.

Test Yourself 1

a Prove that triangle ABC is similar to triangle PQR.

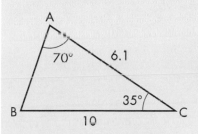 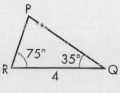

b Find the length of PQ.

Solutions

Test Yourself 1

a Angle ABC = 180 − 70 − 35 = 75
Angle RPQ = 180 − 75 − 35 = 70
Triangles are similar as corresponding angles are equal.

b $\dfrac{PQ}{4} = \dfrac{6.1}{10}$

PQ = 2.4 (1 d.p.)

The area and volume of similar shapes

Revised

- For similar shapes:
  - area scale factor = (linear scale factor)^2.
  - volume scale factor = (linear scale factor)^3.

Test Yourself 2

Two similar cylinders have heights 8 cm and 16 cm.

a The smaller cylinder has a volume of 60 cm^3. Find the volume of the larger cylinder.

b Another similar cylinder has a volume of 202.5 cm^3.
Find its height.

Solutions

Test Yourself 2

a Length scale factor = $\dfrac{16}{8} = 2$
Volume = 60 × 2^3 = 480 cm^3

b Length scale factor = $\left(\dfrac{202.5}{60}\right)^{\frac{1}{3}} = 1.5$

Height = 8 × 1.5 = 12 cm

10 Factorising

Simplifying algebraic fractions

- The rules for simplifying algebraic fractions are the same as for numerical fractions.

> **Chief Examiner says**
>
> Only cancel common factors. An error frequently seen is to cancel the x^2 in expressions such as $\dfrac{x^2 + 3x + 2}{x^2 - 1}$. You cannot cancel x^2 because it is not a common factor.

- To simplify an algebraic fraction,
  - first factorise the numerator and the denominator
  - then cancel any common factors.

Test Yourself 1

a $\dfrac{12x^2}{8x}$

b $\dfrac{6x^2y}{3xy^2}$

c $\dfrac{6ab + 4b^2}{2ab + 2b^2}$

Solutions

Test Yourself 1

a $\dfrac{3x}{2}$ — Divide the numerator and the denominator by $4x$.

b $\dfrac{2x}{y}$ — Divide the numerator and the denominator by $3xy$.

c $\dfrac{2b(3a + 2b)}{2b(a + b)} = \dfrac{3a + 2b}{a + b}$

Factorising expressions of the form $x^2 - a^2$ and $4a^2 - 9b^2$

- Expressions of the form $a^2 - b^2$ can be factorised in the form $(a + b)(a - b)$.
- Expressions in this form are called the difference of two squares.

Test Yourself 2

Factorise these.

a $x^2 - 16$

b $2a^2 - 50b^2$

Solutions

Test Yourself 2

a $x^2 - 16 = x^2 - 4^2 = (x + 4)(x - 4)$

b $2a^2 - 50b^2 = 2(a^2 - 25b^2)$ — Look for a common factor.

$= 2[a^2 - (5b)^2]$

$= 2(a + 5b)(a - 5b)$

Factorising quadratics

Revised

- Some quadratic functions can be factorised.
- Quadratic functions of the form $x^2 + bx + c$ can be factorised to the form $(x + p)(x + q)$ where $p + q = b$ and $p \times q = c$.
    - If c is positive then p and q are both positive or both negative.
    - If c is negative then p and q are of opposite sign.
- Quadratic functions of the form $ax^2 + bx + c$ can be factorised to the form $(px + q)(rx + s)$ where $p \times r = a$, $q \times s = c$ and $ps + qr = b$.
    - If c is positive then q and s are both positive or both negative.
    - If c is negative then q and r are of opposite sign.

Test Yourself 3

Factorise these.

a $x^2 - 7x + 10$

b $x^2 + x - 6$

c $3x^2 - 13x + 4$

d $2x^2 - 5x - 7$

Solutions

Test Yourself 3

a Factors of 10 are 1 and 10, 2 and 5, −1 and −10 or −2 and −5.
   −2 and −5 add to −7.
   $x^2 - 7x + 10 = (x - 2)(x - 5)$

b Factors of −6 are 1 and −6, 2 and −3, −1 and 6 or −2 and 3.
   −2 and 3 add to 1.
   $x^2 + x - 6 = (x - 2)(x + 3)$

c There is no common factor.
   Factors of 3 are 1 and 3.
   c is positive so the signs in the brackets are the same.
   b is negative so the factors are $(3x - ...)(x - ...)$.
   Factors of 4 are −1 and −4 or −2 and −2.
   Check these in the brackets until you get −13x.
   $3x^2 - 13x + 4 = (3x - 1)(x - 4)$

d There is no common factor.
   Factors of 2 are 1 and 2.
   c is negative so the signs in the brackets are different.
   So the factors are $(2x + ...)(x - ...)$ or $(2x - ...)(x + ...)$.
   Factors of −7 are 1 and −7 or −1 and 7.
   Check these in the brackets until you get −5x.
   $2x^2 - 5x - 7 = (2x - 7)(x + 1)$

Chief Examiner says

As with all types of factorising, always check for a common factor first. If there is no common factor for all the terms of a quadratic, there will not be a common factor in a bracket. So factors like $(2x + 4)$ are not possible.

Simplifying more complicated algebraic fractions

Revised

- An algebraic fraction may involve quadratic expressions.
- Factorise the numerator and the denominator separately first.

Test Yourself 4

Simplify the following.

$$\frac{x^2 + 3x + 2}{x^2 - 1}$$

Solution

Test Yourself 4

$\dfrac{(x + 1)(x + 2)}{(x + 1)(x - 1)}$ — Factorise the quadratics.

$= \dfrac{x + 2}{x - 1}$ — Cancel the common factor.

11 Three-dimensional geometry

Three-dimensional (3-D) coordinates

Revised

- To describe a point in three dimensions there are three coordinates.
- The *x*- and *y*-axes lie flat with the *z*-axis vertical.

Test Yourself 1

ABCDEFGH is a cuboid.

Write down the coordinates of the following points.

a C

b H

c F

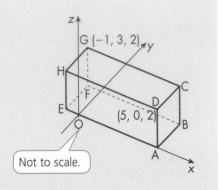

Not to scale.

Solutions

Test Yourself 1

a (5, 3, 2)

b (−1, 0, 2)

c (−1, 3, 0)

Finding lengths and angles in three dimensions

Revised

- The length of the diagonal of a cuboid measuring *a* by *b* by *c* is $\sqrt{a^2 + b^2 + c^2}$.
- You can use right-angled triangles within three-dimensional objects to find lengths and angles.
- See Unit A Chapter 7 for Pythagoras' theorem and Unit A Chapter 14 for the trigonometrical ratios.
- Draw a diagram of the relevant triangle to help you.

ABCDE is a square-based pyramid.
The edge of the base is 30 cm and the height, EH, is 42 cm.

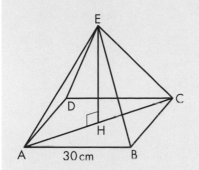 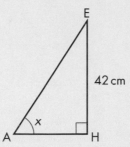

Calculate the following.

a The length of AC

b The angle of EAH

c The length of EA

Test Yourself 2

a $AC^2 = AB^2 + BC^2 = 900 + 900 = 1800$

$AC = \sqrt{1800} = 42.43$ cm (2 d.p.)

b $\tan x = \dfrac{EH}{AH} = \dfrac{42}{21.21...} = 1.98$

Angle EAH = 63.2°

> For accuracy, use all the figures from your answer for AC when finding angle EAH.

c By Pythagoras $\quad EA^2 = EH^2 + AH^2$

$EA^2 = 42^2 + 21.21...^2 = 2213.86$

$EA = 47.1$ cm

> If you kept all the figures in 21.21..., you should find that $21.21...^2 = 450$.

The angle between a line and a plane   Revised

- To find the angle between a line and a plane,
  - draw a perpendicular to the plane from a point on the line
  - join the two lines to make a right-angled triangle
  - use trigonometry.

Find the angle between the line AD and the base of this triangular prism.

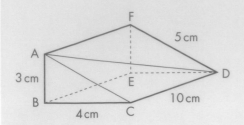

Solution

Test Yourself 3

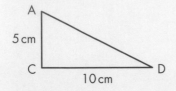

$AD = \sqrt{100 + 25}$

> There are other possible triangles you could use.

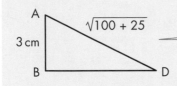

> In this triangle, angle D is the angle between the line and the plane.

Angle $ADB = \sin^{-1} \dfrac{3}{\sqrt{100 + 25}} = 15.6°$

12 Proportion and variation

Revised

- If quantities vary in direct proportion, it means that if you multiply one quantity by a number, you multiply the other by the same number. That is, if you double one quantity, you double the other; if you halve one, you halve the other, and so on.
- You can write the relationship y is proportional to x as $y \propto x$.
- When x and y vary in direct proportion, the graph of y against x is a straight line passing through the origin.

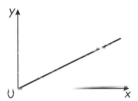

- The gradient of the line, k, is equal to the multiplier.
- The formula for direct proportion is $y = kx$.

Test Yourself 1

The total cost of books is directly proportional to the number of books bought.

If 30 books cost £240, how much does it cost for 120 books?

Solution

Test Yourself 1

$\frac{120}{30} = 4$ — You are given two values of one of the variables. Write these as a fraction to find the multiplier.

$240 \times 4 = £960$

Revised

- In inverse proportion, when one variable increases the other variable decreases.
- Using symbols, y is inversely proportional to x is written as $y \propto \frac{1}{x}$.
- The formula for indirect proportion is $y = \frac{k}{x}$.

Solutions

Test Yourself 2

a The multiplier for speed is $\frac{35}{70} = \frac{1}{2}$, so the multiplier for time is 2.

Time taken $= 40 \times 2 = 80$ minutes or 1 hour 20 minutes.

b The multiplier for speed is $\frac{100}{70} = \frac{10}{7}$, so the multiplier for time is

$\frac{7}{10} = 0.7$.

Time taken $= 40 \times 0.7 = 28$ minutes.

Test Yourself 2

The time taken to travel a given distance is inversely proportional to the speed.

A journey takes 40 minutes at 70 mph.

How long will it take at

a 35 mph

b 100 mph?

Finding formulae

- You may be asked to find the formula for direct proportion or inverse proportion.

a The distance (d miles) travelled at a fixed speed is directly proportional to the time taken (t minutes).

When $d = 200$ miles, $t = 40$ minutes.

Find the distance when the time is 100 minutes.

b The volume, $V\,m^3$, of a given gas is inversely proportional to the pressure $P\,N/m^2$.

When $V = 2\,m^3$, $P = 500\,N/m^2$.

Find the pressure when the volume is $5\,m^3$.

Solutions

Test Yourself 3

a $d \propto t$ or $d = kt$

$200 = 40k$ so $k = 5$

So $d = 5t$

$d = 5 \times 100 = 500$ miles

b $V \propto \dfrac{1}{P}$ or $V = \dfrac{k}{P}$

$2 = \dfrac{k}{500}$ so $k = 1000$

so $V = \dfrac{1000}{P}$

so $5P = 1000$

so $P = \dfrac{1000}{5} = 200\,N/m^2$

Other types of proportion

Revised

- The most common types of direct proportion that you will meet are
  - $y \propto x$
  - $y \propto x^2$
  - $y \propto x^3$
  - $y \propto \sqrt{x}$.
- You may also meet these types of inverse proportion
  - $y \propto \dfrac{1}{x}$
  - $y \propto \dfrac{1}{x^2}$
  - $y \propto \dfrac{1}{\sqrt{x}}$.

The value, £ V, of a diamond is proportional to the square of its mass, W g.

A diamond weighing 14 g is worth £490.

a Find the value of a diamond weighing 40 g.

b Find the mass of a diamond worth £6000.

Solutions

Test Yourself 4

$V \propto W^2$ or $V = kW^2$

$490 = k \times 14^2$

$k = \frac{490}{14^2} = 2.5$

So $V = 2.5 W^2$

a $V = 2.5 \times 40^2$
     $= £4000$

b $6000 = 2.5 W^2$
     $W^2 = 6000 \div 2.5 = 2400$
     $W = \sqrt{2400} = 48.99$ g or 49 g, to the nearest gram

13 Graphs 2

Unit C

Solving simultaneous equations graphically

Revised

- See Unit B Chapter 9 for graphical solution of linear simultaneous equations.

- See Unit C Chapter 5 for graphical solution of a simultaneous equations where one is a quadratic and the other is the equation of a horizontal line ($y = k$).

- You solve simultaneous equations where one is quadratic and the other is any linear equation by the same method as in Chapter 5.

  - Plot the graphs of both equations on the same axes.

  - Read off the coordinates of points where the line cuts the curve. There are usually two solutions.

- If you have the graph of one quadratic equation and are asked to solve a different quadratic equation it may be possible to rearrange the equation rather than draw another graph.

Test Yourself 1

Solve graphically $y = x + 1$ and $y = 2x^2 - 3$.

Solution

Test Yourself 1

x	−1	0	1
$y = x + 1$	0	1	2

x	−2	−1	0	1	2
$y = 2x^2 - 3$	5	−1	−3	−1	5

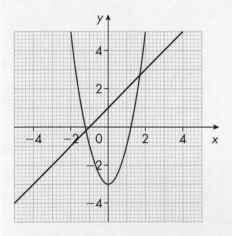

The solutions are $x = 1.7$, $y = 2.7$ and $x = -1.2$, $y = -0.2$.

Test Yourself 2

$x^2 - 3x + 2 = 0$ — Start with the equation you have to solve.

$x^2 - 3x + 3 = 1$

$x^2 - 4x + 3 = 1 - x$ — Do the same thing to both sides until you have the equation of the graph you have been given on one side.

Draw the graph of $y = 1 - x$.

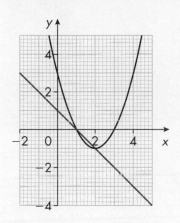

The solutions are $x = 1$ or $x = 2$.

Test Yourself 2

Use the graph of $y = x^2 - 4x + 3$ to solve the equation $x^2 - 3x + 2 = 0$.

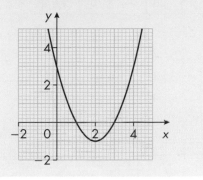

Drawing and recognising other curves   Revised

Cubic graphs

- The cubic graph of $y = ax^3$ has this shape.

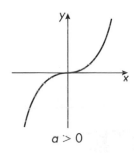

$a > 0$ $\qquad$ $a < 0$

- The general cubic function $ax^3 + bx^2 + cx + d$ has this shape.

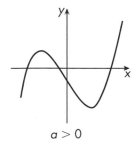

$a > 0$ $\qquad$ $a < 0$

- As with all graphs, the first step in plotting cubic graphs is to make a table of values.

Test Yourself 3

a Draw the graph of $y = x^3 - 3x^2$ for values of x from -2 to 4.

b Use your graph to solve the equation $x^3 - 3x^2 = -1$.

A cubic equation usually has three solutions.

Test Yourself 3

a

x	−2	−1	0	1	2	3	4
y	−20	−4	0	−2	−4	0	16

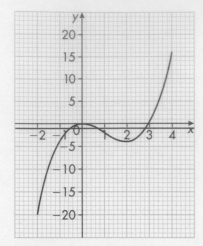

> You may wish to put some extra rows in the table to help you with these calculations.

b   The solution is where the curve crosses the line $y = -1$.
The solution is $x = -0.5, 0.7, 2.9$.

Exponential graphs

- See also Unit C Chapter 8.
- The exponential graph $y = b^x$ has this shape.

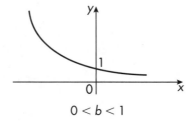

$$b > 0 \qquad\qquad 0 < b < 1$$

- You will only be asked to work out the y values for integer values of x.
- Exponential graphs of the form $y = b^x$ go through the point (0, 1) since $b^0 = 1$ for all positive values of b.
- Exponential graphs of the form $y = Ab^x$ go through the point (0, A).
- When $b > 1$ the curves increase steeply for $x > 0$ and are small when x is negative.
  This is reversed for $0 < b < 1$.

Test Yourself 4

a   Draw the graphs for $y = 3^x$ and $y = \left(\frac{1}{2}\right)^x$ on the same axes.
Use values of x from −3 to +3.

b   Use the your graph to solve the equation $3^x = \left(\frac{1}{2}\right)^x$.

Test Yourself 4

a

x	−3	−2	−1	0	1	2	3
$y = 3^x$	0.04	0.11	0.33	1	3	9	27

x	−3	−2	−1	0	1	2	3
$y = (\frac{1}{2})^x$	8	4	2	1	0.5	0.25	0.125

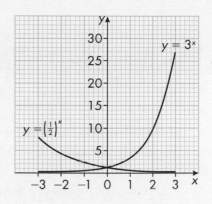

b The solution is the point of intersection of the two curves, $x = 0$, $y = 1$.

So $x = 0$

Reciprocal graphs

- The reciprocal graph $y = \dfrac{a}{x}$ has this shape.

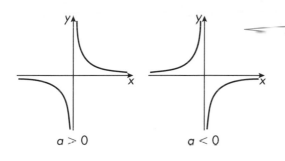

These graphs are in two separate curves. You cannot use 0 as a value for x, since you cannot divide a number by 0.

a Draw the graphs of $y = \dfrac{4}{x}$ and $y = 2x + 1$ on the same axes.

b Use your graph to solve the equation $2x + 1 = \dfrac{4}{x}$.

Test Yourself 5

a

x	−4	−2	−1	1	2	4
$y = \dfrac{4}{x}$	−1	−2	−4	4	2	1

x	−2	0	2
$y = 2x + 1$	−3	1	5

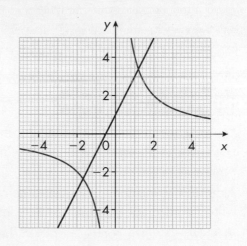

b The solutions are $x = 1.1$ or $x = −1.6$.

14 Quadratic equations

Solving quadratic equations by factorising
Revised

- If the product of two numbers is 0 then one of the numbers must be zero.
- If a quadratic function equals 0 then one or other of its factors equals zero. Equating both factors to zero gives the two solutions to the equation.
- If the quadratic equation is not already factorised, factorise it first. See Unit C Chapter 10 for factorising quadratic expressions.
- If the quadratic expression is not equal to zero, rearrange it before factorising.

Test Yourself 1

Solve these.

a $(3x - 2)(x + 1) = 0$

b $x^2 - 3x = 0$

c $x^2 + x = 6$

Chief Examiner says

- Notice that the sign of the solution is opposite to the sign in the bracket.
- In practical problems the negative solution may be impractical and so should be discarded.

Solutions

Test Yourself 1

a $3x - 2 = 0$ or $x + 1 = 0$

$3x = 2$ or $x = -1$

$x = \frac{2}{3}$ or $x = -1$

$x(x - 3) = 0$ — Factorise.

$x = 0$ or $x - 3 = 0$

$x = 0$ or $x = 3$

b $x^2 + x - 6 = 0$ — Rearrange so that 0 is on one side of the equation.

$(x + 3)(x - 2) = 0$ — Factorise.

$x + 3 = 0$ or $x - 2 = 0$

$x = -3$ or $x = 2$

Completing the square
Revised

- You may be asked to write a quadratic expression of the type $x^2 + bx + c$ in the form $(x + m)^2 + n$. Remember m and n could be positive or negative.
- Alternatively, you could be asked to 'complete the square'. This means write the quadratic expression in the form $(x + m)^2 + n$.
- m is always half the coefficient of x in the quadratic expression.
- The sign in the bracket is always the same as the sign in front of the x term.

Test Yourself 2

Write $x^2 + 10x + 12$ in the form $(x + m)^2 + n$.

Test Yourself 2

$m = 10 \div 2 = 5$

$(x + 5)^2 = x^2 + 10x + 25$ — To make this equal to $x^2 + 10x + 12$ you need to subtract 13.

$x^2 + 10x + 12 = (x + 5)^2 - 13$

Solving quadratic equations by completing the square

Revised

- Make sure the equation has 0 on the right-hand side.
- If the coefficient of x^2 is not 1, divide through the equation by the coefficient of x^2.
- Complete the square on the left-hand side.
- Rearrange the equation to the form $(x + m)^2 = n$.
- Take the square root of both sides, remembering that the square root of a number may be positive or negative.
- If you are asked to give an exact answer, leave your answer in surd form.

Test Yourself 3

Solve $2x^2 - 12x + 1 = 0$
by completing the square.

Solution

Test Yourself 3

$x^2 - 6x + 0.5 = 0$ — Divide both sides by 2.

$(x - 3)^2 - 3^2 + 0.5 = 0$ — Complete the square.

$(x - 3)^2 - 8.5 = 0$

$(x - 3)^2 = 8.5$

$(x - 3) = \pm\sqrt{8.5}$

$x = 3 \pm \sqrt{8.5}$ — This is the exact answer.

$= 5.92$ or 0.08 correct to 2 decimal places.

Solving quadratic equations by using the formula

Revised

- The solutions to the equation $ax^2 + bx + c = 0$ are given by the formula $x = \dfrac{-b \pm \sqrt{b^2 - 4ac}}{2a}$
- This formula is given on the examination paper. Take care with signs and always work out the numerator before dividing by $2a$.
- Use the formula when the expression does not factorise. If you are asked to round your answers then it is unlikely that the expression will factorise.

Test Yourself 4

Solve $2x^2 - 7x - 3 = 0$ by using the formula.

Solution

Test Yourself 4

$a = 2, b = -7, c = -3$

$x = \dfrac{-(-7) + \sqrt{(-7)^2 - 4 \times 2 \times (-3)}}{2 \times 2}$

$= \dfrac{7 \pm \sqrt{49 + 24}}{4}$

$= \dfrac{7 + \sqrt{73}}{4}$ or $\dfrac{7 - \sqrt{73}}{4}$ — Press '=' before dividing by 4.

$= 3.89$ or -0.39 correct to 2 decimal places.

15 Simultaneous equations

Solving simultaneous equations by substitution

- When solving simultaneous equations you are looking for a single value of x and a single value of y which satisfy both equations.
- See Unit B Chapter 9 for solving simultaneous equations by the elimination method.
- The substitution method is the best method if y is the subject of one or both of the equations.

  Step 1: Substitute the equation with y as the subject into the other equation.

  Step 2: Solve this equation to find the value of x.

  Step 3: Substitute this value into the equation with y as the subject to find the value of y.

- If x is the subject of one of the equations the method can be reversed by substituting for x.

Test Yourself 1

Use the substitution method to solve

$2x + 3y = 10$ (1)

$y = 2x + 6$ (2)

Solution

Test Yourself 1

$2x + 3(2x + 6) = 10$ — Substitute (2) in (1). Don't forget the brackets.

$2x + 6x + 18 = 10$

$8x = -8$

$x = -1$

$y = 2 \times (-1) + 6 = 4$

Solution $x = -1$, $y = 4$ — Substitute $x = -1$ in (2).

Solving one linear and one quadratic equation

- See Unit C Chapter 13 for solving one linear and one quadratic equation graphically. Remember that solving equations graphically does not give an exact answer.
- If both equations are in the form '$y =$', then eliminate y by equating them. Otherwise rearrange the linear equation so that y is the subject, then substitute for y in the quadratic equation.
- Simplify and rearrange the resulting quadratic equation so that one side is zero.
- Solve the quadratic equation. See Unit C Chapter 14 for methods for solving quadratic equations.
- Substitute each value of x in the linear equation to find the corresponding value for y.

Chief Examiner says

If you are asked to solve two simultaneous equations give the solution as two ordered pairs, each x value with the corresponding y value. If you are asked to find the intersection of two graphs, give the answers as coordinates.

Test Yourself 2

$$x^2 + 4x - 1 = 3x + 1$$
$$x^2 + x - 2 = 0$$
$$(x + 2)(x - 1) = 0$$

So $x + 2 = 0$ or $x - 1 = 0$
$$x = -2 \text{ or } x = 1$$

When $x = -2$, $y = 3 \times (-2) + 1 = -6 + 1 = -5$
When $x = 1$, $y = 3 \times 1 + 1 = 4$

So the graphs intersect at $(-2, -5)$ and $(1, 4)$.

Test Yourself 3

$$x + x^2 - 2x - 2 = 1$$ ← Substitute for y in the linear equation.

$$x^2 - x - 3 = 0$$ ← Rearrange so that the right-hand side is equal to zero.

$$x = \frac{1 \pm \sqrt{1 + 4 \times 3}}{2}$$ ← Solve using the quadratic equation.

$$x = 2.30 \text{ or } x = -1.30$$

When $x = 2.30$, $y = 2.30^2 - 2 \times 2.30 - 2 = 1.31$
When $x = -1.30$, $y = -1.30^2 - 2 \times -1.30 - 2 = 2.29$

So solutions are $x = 2.30$, $y = -1.31$ and $x = -1.30$, $y = 2.29$

Test Yourself 2

Find algebraically the coordinates of the intersection of the line $y = 3x + 1$ and the curve $y = x^2 + 4x - 1$.

Test Yourself 3

Solve these simultaneous equations by substitution.
$$x + y = 1$$
$$y = x^2 - 2x - 2$$

16 Trigonometry

The area of a triangle, The sine rule and The cosine rule

- This is the convention for labelling the sides and angles of a triangle.
- The area of the triangle is

$$\text{Area} = \tfrac{1}{2} ab \sin C$$
$$= \tfrac{1}{2} bc \sin A$$
$$= \tfrac{1}{2} ac \sin B$$

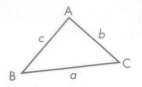

- The sine rule is

$$\frac{\sin A}{a} = \frac{\sin B}{b} = \frac{\sin C}{c}$$

$$\frac{a}{\sin A} = \frac{b}{\sin B} = \frac{c}{\sin C}$$

- The cosine rule is

$$a^2 = b^2 + c^2 - 2bc \cos A$$
$$b^2 = c^2 + a^2 - 2ca \cos B$$
$$c^2 = a^2 + b^2 - 2ab \cos C$$

Chief Examiner says

When you are asked to find lengths or angles in a triangle without a right-angle, you need to use the sine or cosine rule.

Test Yourself 1

a Find the lengths of AC and AB.

b Calculate the area of triangle ABC.

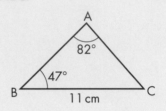

Solutions

Test Yourself 1

a $\dfrac{a}{\sin A} = \dfrac{b}{\sin B}$

$\dfrac{11}{\sin 82°} = \dfrac{b}{\sin 47°}$

$b = \dfrac{11 \times \sin 47°}{\sin 82°} = 8.12\,\text{cm}$

$AC = 8.12\,\text{cm}$

Angle ACB
$= 180° - 47° - 82° = 51°$

$\dfrac{c}{\sin C} = \dfrac{a}{\sin A}$

$c = \dfrac{11 \times \sin 51°}{\sin 82°} = 8.63\,\text{cm}$

$AB = 8.63\,\text{cm}$

b Using $\tfrac{1}{2} ac \sin B$,

Area
$= \tfrac{1}{2} \times 11 \times 8.63 \times \sin 47°$
$= 34.7\,\text{cm}^2$

Graphs of trigonometrical functions

- You need to know the shape of the graphs of $y = \sin x$ and $y = \cos x$.
  - They are both repeating wave shapes. They repeat every 360°. The length of a repeating pattern like this is called the period of the graph.
  - The height of the wave above the mean is 1 unit. This height is called the amplitude of the wave.
  - The symmetries of these graphs may be used to find other angles which have the same sin or cos value. For instance
  $\sin 50° = \sin 130° = \sin 410°$ and $\cos 100° = -\cos 80°$

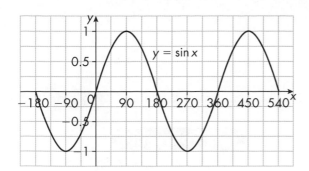
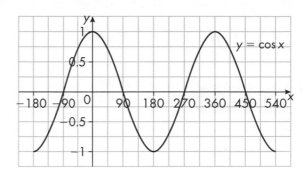

- You also need to know the shape of the graph of $y = \tan x$.
  - It is not a continuous curve.
  - Its period is 180°.
  - The broken lines are called asymptotes.

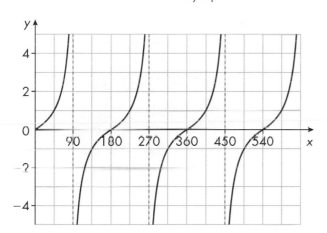

Test Yourself 2

a Use the graphs above to estimate the value of these.

   i   sin 120°

   ii cos 200°

b Use the graphs above to find four approximate solutions to the following.

   i   sin $x = 0.25$

   ii cos $x = 0.75$

Solutions

Test Yourself 2
a i 0.9
   ii −0.9
b i 15°, 165°, 375°, 525°
   ii −40°, 40°, 320°, 400°

Other trigonometrical functions   Revised

- The graphs of $y = A \cos Bx$ and $y = A \sin Bx$ each have amplitude A and period $\dfrac{360}{B}$.

Solutions

Test Yourself 3
a $y = 2 \sin 3x$
b $y = 1.5 \cos \frac{1}{2}x$

Test Yourself 3

Write down the equations of these two curves.

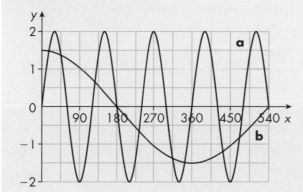

17 Functions

Function notation Revised

- See Unit A Chapter 16 for linear functions.
- You deal with quadratic functions in the same way.

Solutions

Test Yourself 1

a $f(4) = 4^2 - 3 \times 4 + 2 = 6$

b $f(-2) = (-2)^2 - 3 \times (-2) + 2 = 12$

Test Yourself 1

If $f(x) = x^2 - 3x + 2$, find the values of these.

a $f(4)$

b $f(-2)$

Translations Revised

- The graph of $y = f(x) + a$ is the graph of $y = f(x)$ translated by $\begin{pmatrix} 0 \\ a \end{pmatrix}$.

- The graph of $y = f(x + a)$ is the graph of $y = f(x)$ translated by $\begin{pmatrix} -a \\ 0 \end{pmatrix}$.

> **Chief Examiner says**
>
> There is no need to scale the axes for a sketch graph. However, you should show important features, such as where the graph crosses the axes.

Test Yourself 2

The graph shows $y = x^2$.
On a copy of this graph, sketch $y = x^2 + 2$.

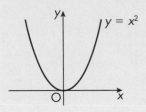

Solution

Test Yourself 2

$y = x^2 + 2$ is the same shape as

$y = x^2$ translated by $\begin{pmatrix} 0 \\ 2 \end{pmatrix}$.

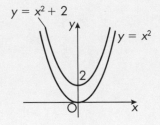

One-way stretches Revised

- The graph of $y = kf(x)$ is the graph of $y = f(x)$ stretched in the y direction with scale factor k.
- The graph of $y = f(kx)$ is the graph of $y = f(x)$ stretched in the x direction with scale factor $\frac{1}{k}$.

Test Yourself 3

The graph shows
$y = \sin x$.

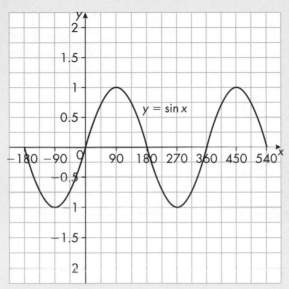

On separate copies of the graph, sketch graphs of these functions.

a $y = 2 \sin x$

b $y = \sin 2x$

Solutions

Test Yourself 3

a The graph of $y = 2 \sin x$ is the graph of
$y = \sin x$ stretched in the y direction with
scale factor 2.

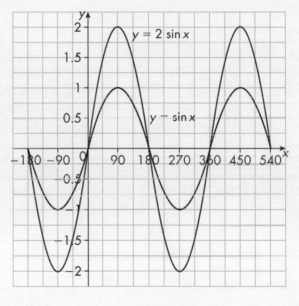

b The graph of $y = \sin 2x$ is the graph of $y = \sin x$ stretched in the x direction with scale factor $\frac{1}{2}$.

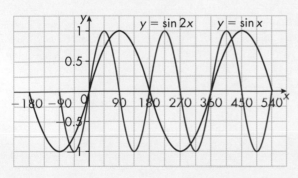

Unit C   117

Reflections

- Reflections are one-way stretches with a scale factor of -1.
- The graph of $y = f(-x)$ is the graph of $y = f(x)$ reflected in the y-axis.
- The graph of $y = -f(x)$ is the graph of $y = f(x)$ reflected in the x-axis.

Solution

Test Yourself 4

If $f(x) = x^2 - 3x$ then $-f(x) = 3x - x^2$

So the graph of $y = 3x - x^2$ is the reflection of $y = x^2 - 3x$ in the x-axis.

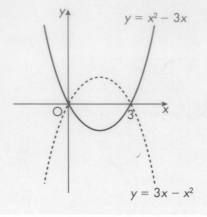

Test Yourself 4

The graph shows $y = x^2 - 3x$.

On a copy of the graph sketch the graph of $y = 3x - x^2$.

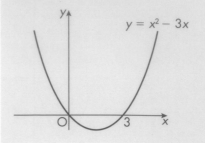

18 Length, area and volume

Arcs and sectors Revised

- See Unit B Chapter 11 for circle terms.

- Arc length $= \dfrac{\theta}{360} \times 2\pi r$

- Area of sector $= \dfrac{\theta}{360} \times \pi r^2$

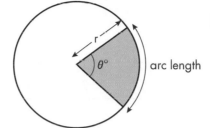

Chief Examiner says

Sector area is directly proportional to the angle at the centre.

Test Yourself 1

Find the arc length and the area of this sector.

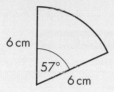

6 cm

57°

6 cm

Solution

Test Yourself 1

Arc length $= \frac{57}{360} \times 2 \times \pi \times 6 = 5.97$ cm

Sector area $= \frac{57}{360} \times \pi \times 6^2 = 17.91$ cm²

Pyramids, cones and spheres Revised

- For a pyramid, the shape of the base is usually part of its name.
  For example 'a square-based pyramid'.
  Volume, $V = \frac{1}{3} \times$ area of base $\times$ height

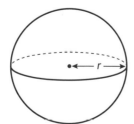

- A 'pyramid' with a circular base is a cone.
  Volume, $V = \frac{1}{3}\pi r^2 h$
  Curved surface area, $A = \pi r l$

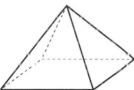

Test Yourself 2

a A pyramid has a square base of side 4 m and is 9 m high. Calculate the volume of the pyramid.

b Find the height of a cone with volume 2.5 litres and base radius 10 cm.

c A sphere has a surface area of 500 cm². Calculate the radius of the sphere.

- For a sphere,
  $V = \frac{4}{3}\pi r^3$
  Surface area $= 4\pi r^2$

Test Yourself 2

a $V = \frac{1}{3} \times (4 \times 4) \times 9$

$\quad = 48 \text{ m}^3$

b $2500 = \frac{1}{3} \times \pi \times 10^2 \times h$

$\quad h = \dfrac{3 \times 2500}{\pi \times 100}$ ◁─── Remember that 1 litre = 1000 cm³.

$\quad h = 23.9 \text{ cm}$

c $500 = 4\pi r^2$

$\quad r^2 = \dfrac{500}{4\pi}$

$\quad r = 6.3 \text{ cm}$

Chief Examiner says

Don't forget that $4\pi r^2$ is 4π times r^2, not $(4\pi r)^2$.

Compound shapes and problems

Revised ☐

- You need to be able to work with complex shapes such as the segment of a circle and the frustum of a cone.
- Volume of a frustum = Volume of whole cone − Volume of missing cone.

Solution

Test Yourself 3

Area of segment = Area of sector − Area of triangle

Segment $\quad = \text{sector} - \text{triangle}$

$\quad = \frac{104}{360} \times \pi \times 8^2 - \frac{1}{2} \times 8 \times 8 \times \sin 104°$

$\quad = 27.0 \text{ cm}^2$

Test Yourself 3

Find the area of the shaded segment.

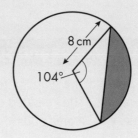

19 Probability 2

The addition rule and The multiplication rule

- See Unit C Chapter 4 for more on probability.
- Equally likely outcomes may be listed in a table or shown on a grid.
- Two events are mutually exclusive if they cannot both occur at the same time.
- For mutually exclusive events $P(A \text{ or } B) = P(A) + P(B)$.
- Two events are independent when the outcome of the second event is not affected by the outcome of the first.
- For independent events $P(A \text{ and } B) = P(A) \times P(B)$.

Chief Examiner says

- When listing possibilities, be systematic to make sure you don't miss any possibilities.
- One of the most common errors is to add probabilities instead of multiplying them. If you see the word 'and' or 'both' or 'all' then multiply the probabilities.
- If you are finding the probability of all but one of the outcomes, it is often easier to work out $1 - P(\text{remaining outcome})$.

Probability tree diagrams

- These are useful when dealing with two or more events.
- Each set of branches shows the possible outcomes of the event.
- The probabilities on each set of branches should add to 1.
- To find the probabilities of the combined events, multiply the probabilities along the branches.

Test Yourself 1

This tree diagram shows the probabilities that Penny has to stop at lights or a level crossing on her way to work.

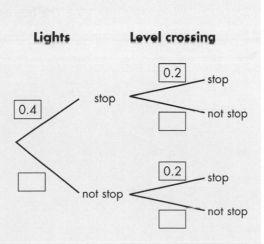

a Complete the tree diagram.

b What is the probability that

   i   she does not stop at either the lights or the level crossing

   ii she stops at either the lights or the level crossing but not both?

Solutions

Test Yourself 1

a

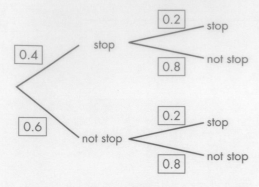

Lights   Level crossing

b i   P(does not stop at either) = 0.6 × 0.8 = 0.48

  ii   P(stops at either but not both) = 0.4 × 0.8 + 0.6 × 0.2

                                       = 0.32 + 0.12 = 0.44

> **Chief Examiner says**
>
> Even if it is not asked for, it is a good idea to draw a tree diagram.

Dependent events

Revised ☐

- When events are not independent, the outcome of one affects the probability that the other happens.
- Tree diagrams are very useful for dealing with dependent events.
- In a tree diagram in this case, the probabilities on the second pairs of branches will be different.

Solution

Test Yourself 2

Wakes   **Misses bus**

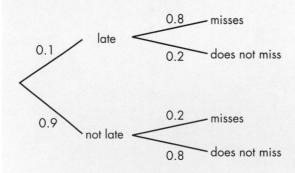

P(miss bus) = 0.1 × 0.8 + 0.9 × 0.2

               = 0.08 + 0.18 = 0.26

> **Test Yourself 2**
>
> The probability that Pali wakes up late on a work morning is 0.1. When he wakes up late, the probability that he misses the bus is 0.8. When he doesn't wake up late, the probability that he misses the bus is 0.2. Draw a tree diagram to represent this, and find the probability that he misses the bus on a work morning.

> **Chief Examiner says**
>
> Remember that, at each stage, the sum of the probabilities is 1.

20 Algebraic fractions

- See Unit C Chapter 10 for simplifying algebraic fractions.

Adding and subtracting algebraic fractions ———————— Revised ☐

- The rules for adding and subtracting algebraic fractions are the same as for numerical fractions.
- First write the fractions with a common denominator. Then add or subtract the numerators.

Solutions

Test Yourself 1

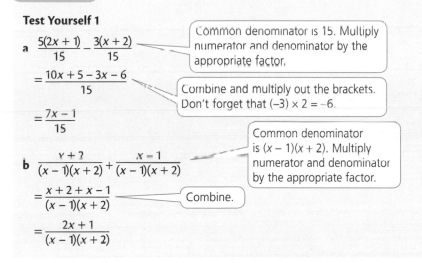

a $\dfrac{5(2x + 1)}{15} - \dfrac{3(x + 2)}{15}$ — Common denominator is 15. Multiply numerator and denominator by the appropriate factor.

$= \dfrac{10x + 5 - 3x - 6}{15}$ — Combine and multiply out the brackets. Don't forget that $(-3) \times 2 = -6$.

$= \dfrac{7x - 1}{15}$

b $\dfrac{x + 2}{(x - 1)(x + 2)} + \dfrac{x - 1}{(x - 1)(x + 2)}$ — Common denominator is $(x - 1)(x + 2)$. Multiply numerator and denominator by the appropriate factor.

$= \dfrac{x + 2 + x - 1}{(x - 1)(x + 2)}$ — Combine.

$= \dfrac{2x + 1}{(x - 1)(x + 2)}$

> **Test Yourself 1**
>
> Simplify the following.
>
> **a** $\dfrac{2x + 1}{3} - \dfrac{x + 2}{5}$
>
> **b** $\dfrac{1}{x - 1} + \dfrac{1}{x + 2}$

Solving equations involving algebraic fractions ———————— Revised ☐

- Multiply both sides by the common denominator.

Solutions

Test Yourself 2

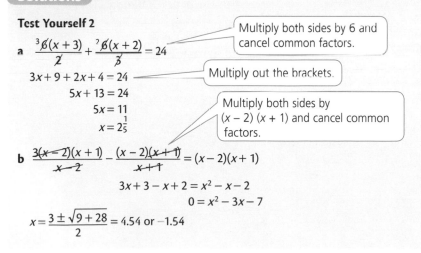

a $\dfrac{^3\cancel{6}(x + 3)}{\cancel{2}} + \dfrac{^2\cancel{6}(x + 2)}{\cancel{3}} = 24$ — Multiply both sides by 6 and cancel common factors.

$3x + 9 + 2x + 4 = 24$ — Multiply out the brackets.

$5x + 13 = 24$

$5x = 11$ — Multiply both sides by $(x - 2)(x + 1)$ and cancel common factors.

$x = 2\tfrac{1}{5}$

b $\dfrac{3(\cancel{x - 2})(x + 1)}{\cancel{x - 2}} - \dfrac{(x - 2)(\cancel{x + 1})}{\cancel{x + 1}} = (x - 2)(x + 1)$

$3x + 3 - x + 2 = x^2 - x - 2$

$0 = x^2 - 3x - 7$

$x = \dfrac{3 \pm \sqrt{9 + 28}}{2} = 4.54 \text{ or } -1.54$

> **Test Yourself 2**
>
> Solve these.
>
> **a** $\dfrac{x + 3}{2} + \dfrac{x + 2}{3} = 4$
>
> **b** $\dfrac{3}{x - 2} - \dfrac{1}{x + 1} = 1$
>
> This could be written as
> $\tfrac{1}{2}(x + 3) + \tfrac{1}{3}(x + 2) = 4$